UNDERSTANDING POPULATION GENETICS

UNDERSTANDING POPULATION GENETICS

THROUGH THE DERIVATION OF TEN MAJOR RESULTS

TORBJÖRN SÄLL AND
BENGT O. BENGTSSON

Registered Office:
John Wiley & Sons, Ltd, The Atrium, Southern Gate, Chichester, West Sussex, PO19 8SQ, UK

Editorial Offices:
9600 Garsington Road, Oxford, OX4 2DQ, UK
The Atrium, Southern Gate, Chichester, West Sussex, PO19 8SQ, UK
111 River Street, Hoboken, NJ 07030-5774, USA

For details of our global editorial offices, for customer services and for information about how to apply for permission to reuse the copyright material in this book please see our website at www.wiley.com/wiley-blackwell.

Library of Congress Cataloging-in-Publication Data

Names: Säll, T. (Torbjörn), 1957- author. | Bengtsson, Bengt O. 1946-author.
Title: Understanding population genetics / Torbjörn Säll, Bengt O. Bengtsson.
Description: Chichester, West Sussex ; Hoboken, NJ : Wiley-Blackwell, 2017. |
    Includes bibliographical references and index. |
Identifiers: LCCN 2017016746 (print) | LCCN 2017018822 (ebook) | ISBN
    9781119124078 (pdf) | ISBN 9781119124054 (epub) | ISBN 9781119124030
    (paperback)
Subjects: | MESH: Genetics, Population | Problems and Exercises
Classification: LCC QH455 (ebook) | LCC QH455 (print) | NLM QU 18.2 | DDC
    576.5/8–dc23
LC record available at https://lccn.loc.gov/2017016746

A catalogue record for this book is available from the British Library.

Cover Design: Wiley
Cover Image: © Lava4images/Getty Images

Set in 10/13pt ACaslonPro by SPi Global, Chennai, India

Printed in the UK

In science nothing capable of proof
ought to be accepted without proof.

R. Dedekind, 1888

CONTENTS

Introduction: On the constancy of allele frequencies

The most important result in population genetics needs no mathematics to be understood: If in a population an allele is unaffected by any systematic evolutionary forces such as mutation or selection, then its relative frequency is expected to stay the same over generations.

This result is generally called the Law of Constancy of Allele Frequencies, and to our first-year students we often teach it as: 'If nothing happens, nothing happens'. It follows from the 'hard' nature of genes: the homozygotes produce gametes of their own kind, while heterozygotes produce gametes with the two types in equal proportion – just as they themselves once were formed. The transmission of genes between generations implies, in itself, no change in the frequencies of their different kinds. Even if this result is simple, it is not trivial. It means, for example, that the spontaneous idea – which can be found from the early days of Mendelism up till today – that dominant alleles always win in evolution over recessive alleles is mistaken.

Thus, mathematics is not needed to grasp the most fundamental result in population genetics, but it certainly helps when a more detailed understanding is called for. The argument given above is excellent as a first approximation. We know, however, that mutations *do* occur and that survival and fertility *do* differ between genotypes. So, what happens when these factors are taken into account? Without any deeper scrutiny of the problem, it can be predicted that as long as mutations are rare and selection isn't strong, allele frequencies will change only slowly. But what do 'rare' and 'strong' and 'slowly' mean here? Only a more structured, quantitative analysis will tell us.

Understanding Population Genetics, First Edition. Torbjörn Säll and Bengt O. Bengtsson.
© 2017 John Wiley & Sons Ltd. Published 2017 by John Wiley & Sons Ltd.

A certain amount of mathematics is therefore necessary for any deeper understanding of population genetics. The methods needed are not very advanced; anyone studying natural science at university level will know them from school, even when the precise details have been forgotten. This can be seen from the way in which elementary population genetics is normally taught. In lectures and textbooks, phrases such as 'it can be shown that ...' or 'it is easily proven that ...' abound, thereby making life simpler for students as well as teachers, since time is not spent on routine calculations. The emphasis is on speed, and no one particularly likes devoting, say, half an hour to a detailed derivation. But perhaps something important is lost in this pedagogical streamlining?

This book is written out of our belief that to reach a deeper understanding of population genetics, it is necessary to know how the most important results can be derived. With all details included and in the most concrete manner. To us, population genetics resembles other specialized genetic techniques: to learn how they function and what they can be used for, direct hands-on experience is essential. If one really wants to know what it is all about, one must get in there and get one's hands dirty.

It is our goal to help the reader obtain this knowledge, and for this purpose the book presents detailed derivations of ten important results. Since proofs may be analysed at different levels of ambition, we present the derivations so that the successive steps can be carefully followed or just glanced over, depending on the interests and needs of the reader. Throughout the text, we try to be close to the reader, helping with every potential difficulty.

Special attention is given to the assumptions that underlie the results. Not only is this important in a text that aims at presenting a logical flow of argument, it is also important at a deeper level. The results we present are certainly interesting – and sometimes even beautiful. But they first come to life when they are adapted to concrete situations by scientists who wish to understand what actually goes on in nature.

How empirical data and theoretical results combine to produce knowledge about the world is, however, never a self-evident affair, and much can be written about it. Here, we just wish to comment on one of its aspects. We are convinced that for such knowledge to be produced, the scientist involved must have a clear feeling for what the theoretical results actually mean – a feeling that is neither too limited, nor too vague. An example: The arguments normally used to reach the so called Hardy–Weinberg proportions of genotypes restrict the result to infinitely large random-mating populations – that is, situations which never exist in real life. But many actual populations are very large in number, and the individuals in them mate without any obvious pattern; this is why scientists often take on the responsibility to disregard the exact assumptions used to reach the Hardy–Weinberg proportions and assume that these proportions actually hold for the population they investigate. And they do so quite rightly, since they know how relatively unimportant the effects will be that follow from the small deviations to the theoretical assumptions.

Any scientist who wants to apply a theoretical result to real-world data must, in a similar way, know when to ignore the discrepancies between the theory and the functioning of the actual word – and when *not* to apply the theory to any particular case. Experience helps when it comes to making such decisions, but so too does a sound knowledge of the exact conditions that go into the theoretical conclusions and how severe the effects will be if these assumptions are modified. Detailed proof-analysis is therefore a fruitful path towards understanding how the real world actually functions.

Who will gain from obtaining the improved understanding of population genetics that comes from working through the derivations and questions presented here?

In our experience, the answer is: more people than ever before. During its first one hundred years, theoretical population genetics was a rather arcane speciality among many other branches of genetics. Over time, it developed an impressive body of interesting results, of which at least some were of direct relevance to, among others, evolutionary biologists and clinical geneticists. But just as Darwin could argue for evolution by natural selection without knowing any genetics (and even less population genetics), most practitioners of genetics could cope without any deeper knowledge of the field.

[3]

With the development of modern genomics, the situation has radically changed. Today, almost every medical or biological researcher is inundated by masses of detailed genetic data. To figure out what they mean, the standard procedure is to feed the data into special and well-designed computer programs which (perhaps after some initial parameter-setting by the researcher) produce lists of estimates and probability values. Well, what do *these* mean? Can the values be trusted? What underlying assumptions went into *their* calculations? Are the estimates obtained weak or strong? And what do all the names on the different estimates stand for?

Many biology and medical researchers are today asking themselves these questions, whether they are studying the genetic background to a particular disease, the phylogenetic relationship between some organisms or the evolution of an ecologically important trait. A deeper understanding of population genetics becomes, then, of direct help; in particular, since it is often results from theoretical population genetics that underlie the formulas used by the handy computer programs.

The key results that we analyse here in detail have been chosen to reflect the scope and breadth of the whole of biology. We have tried to present the derivations so that they become as informative as possible in their balance between the simple and the complex. Some of the chapters deal with formulae that describe how the genetic composition of a population changes over time as a function of – for example – mutation, genetic drift or selection. Other chapters describe key concepts in population genetics, such as linkage disequilibrium, coalescence time or the geographic substructure measure F_{ST}, and show why they are useful in the mapping of nature. Our focus throughout is on reaching key results, that is formulae which in a striking form summarize deductive conclusions based on abstract and well-defined models of reality. These results are often major achievements with interesting intellectual histories (here only hinted at), but also with productive futures, since further analyses and results are continually built on their basis.

With the accompanying broad discussions – often extending far beyond the chapter headings – these ten derivations therefore give an overview of the usefulness, versatility and scientific value that comes with the logical precision of theoretical population genetics. Together they should provide a solid foundation on which more advanced studies can be based. After having worked through the book, readers will better grasp what computer programs analysing sequence data do (and don't do), the assumptions that underlay some advanced evolutionary arguments, and the logic and limitations of phylogenetic reconstructions. The tools for understanding the research literature will be at hand, and it should be possible – for those that are interested – to develop new ideas in the field.

In this introduction, we have used terms such as 'gametes', 'genes', 'alleles', 'mutation', 'selection', 'Hardy–Weinberg proportions' and 'genetic drift', and we assume that the reader is (reasonably) familiar with them and feels (reasonably) comfortable with their use. If you are not, find a standard text on genetics and read it before you continue. Some of these terms are also included in the Glossary at the end of the book. Words and terms discussed there are normally written in italics when first introduced in the derivations. Wherever possible, we have tried to be generous with explanations.

When it comes to the mathematics, everything more advanced than straightforward high-school techniques is explained and/or discussed along the way   see, in particular, the sections named 'Background'. The amount of detail that we devote to the derivations may be boring to more experienced readers, but such parts can be easily skipped. The book is meant to be read differently by every reader: some may read it slowly and try to follow every step in the argument, others will concentrate on specific chapters and try to grasp the wider implications of the results obtained there. Only a small number of references are given. The interested reader can easily find theoretical extensions and illustrative empirical examples by searching the Internet. Some of the references are included to situate population genetics as the historical adventure it is.

The derivations have been taught by us in various courses. The basic idea of the book is, however, that it should be so self-contained that it can be read by anyone *without* any lectures or help from any teacher. We have often met students and colleagues who wished that they could learn some more population genetics on their own or in a small group of friends, so that they could better tackle the problems they encountered in their research. This book is for them.

Finally: Understanding population genetics is fun. Knowing how complex data can be fruitfully condensed and how reasonable predictions can be made about future populations is fun. Experiencing the beauty of a nice theoretical result emerging out of a complicated mess of calculations is fun. For us, writing this book has been great fun. We hope that some of these joyful feelings will be experienced by the reader.

Derivation 1

Balance between forward and backward mutation

The Law of Constancy of Allele Frequencies is described in the Introduction, and in our first derivation we investigate how it is affected by mutation. The absolute fixity of alleles is therefore no longer taken for granted, since they are now allowed to change into one another. The process this generates is followed over time in a well-defined model.

The derivation leads to a differential equation that shows how rapidly the allele frequencies change and in what direction. Since mutation normally is a weak force, other processes – such as genetic drift – may also affect these frequencies. The complications this leads to are discussed at the end of this chapter, where we put the derived results in a wider perspective.

Analysis

Assumptions and definitions

Consider a well-delimited population of an organism. In the organism there is a gene, A, for which only two allelic states are possible, A_1 and A_2. Let p_t be the relative frequency of A_1 in the present generation, which we denote with index t. If this frequency is known, so is the frequency of allele A_2 (it is always equal to $1 - p_t$). There is no selection favouring either of the two alleles – they are what is called *selectively neutral*. Let μ_{12} be the probability that any given A_1 allele mutates to A_2 during the span of a single generation, and μ_{21} the probability that any given A_2 allele mutates to A_1; both values are assumed to be small but greater than 0.

Understanding Population Genetics, First Edition. Torbjörn Säll and Bengt O. Bengtsson.
© 2017 John Wiley & Sons Ltd. Published 2017 by John Wiley & Sons Ltd.

(A list of Greek letters commonly used in population genetics is given in Background 1:1.) Finally, we assume that the population size is very large.

The fact that mutation can occur in both directions makes it reasonable to believe that the population will evolve towards a *polymorphism* where the allele frequency p stays the same in generation after generation. We will investigate whether this is the case and find the exact value for the allele frequency at this *equilibrium*. Furthermore, we will derive an expression that describes how p changes over time with respect to its equilibrium value.

The recursion system and its equilibrium

Given these assumptions and definitions, we can express the value of p in generation $t + 1$ as a function of the mutation rates and the value of p in generation t. The following logic is used: the A_1 alleles in generation $t + 1$ are the ones that were A_1 in the previous generation and did not mutate, plus those that were A_2 and mutated to A_1. In mathematical terms, this can be written:

$$p_{t+1} = p_t(1 - \mu_{12}) + (1 - p_t)\mu_{21}. \tag{1.1}$$

By assuming that the population is very large, we can ignore that the allele frequency in generation $t + 1$ is also affected by chance effects arising from the sampling of parents and gametes in generation t. The role of the so-called *genetic drift* becomes smaller with the magnitude of the population size – just like the relative deviation from 1:1 becomes smaller when a fair coin is tossed an increasing number of times. In Derivation 6, the relationship between population size and genetic drift is discussed in more detail.

Expression (1.1) describes a *recursion system*; that is, it gives a mathematically formulated description of how a system at a certain time depends on the state of the system one unit of time earlier (plus a number of parameters that do not change with time). Almost all of population genetics builds on such recursive relationships. They are, by far, the best tools we have for investigating evolutionary processes, since evolution is, in population genetic terms, nothing other than changes in allele frequencies.

The first step in the analysis of expression (1.1) is to investigate whether there is an equilibrium value for p such that the mutations in opposite direction balance and the frequency p stays constant over time. At such an equilibrium, it must hold that $p_{t+1} = p_t = p_{eq}$. For the purpose of finding this value, we substitute p_{eq} into (1.1), which gives:

$$p_{eq} = p_{eq}(1 - \mu_{12}) + (1 - p_{eq})\mu_{21}. \tag{1.2}$$

We can solve for p_{eq} as follows:

$$p_{eq} = p_{eq} - p_{eq}\mu_{12} + \mu_{21} - p_{eq}\mu_{21}$$

$$0 = -p_{eq}\mu_{12} + \mu_{21} - p_{eq}\mu_{21}$$

$$p_{eq}\mu_{12} + p_{eq}\mu_{21} = \mu_{21}$$

$$p_{eq}(\mu_{12} + \mu_{21}) = \mu_{21}$$

$$p_{eq} = \mu_{21}/(\mu_{12} + \mu_{21}) \tag{1.3}$$

Since both μ_{12} and μ_{21} are positive, it follows that equation (1.1) has exactly one equilibrium point and that the equilibrium value, p_{eq}, is greater than 0 and smaller than 1, which is necessary for the result to be valid in our context. The first explicit derivation of this elementary formula seems to be due to Wright (1931). It is self-evident that if allele A_1 moves to an equilibrium where its frequency is p_{eq}, then allele A_2 will move to an equilibrium with frequency $1 - p_{eq}$.

(When, in this book, we write that 'x is greater than y' or that 'z is positive', we conform to everyday language and mean than $x > y$ or that $z > 0$, different from the mathematical practice which here would imply a $\geq$ sign instead. In general, we are not interested in what happens in 'special cases' – situations where there is an exact functional relationship between key parameters – since these are normally of limited population genetic interest.)

The change in p over time

Our next task is to find an expression for how p changes with time, given the mutation rates and an assumed initial value for p that we will denote p_0. For this purpose, we develop expression (1.1) into a differential equation and then use the results in Background 1:2 to solve it.

From a difference equation to a differential equation

Expression (1.1) is what is called a *difference equation*; it describes how a dynamic process evolves in a step-wise fashion. Such equations are very common in population genetics (and very easy and convenient to study with computer simulations). Their great advantage is that they give a clear view of what happens with respect to the genetic material when one generation produces the next. They normally lack any explicit solution, however, and their mathematical behaviour may turn chaotic when 'pushed' with strong parameter values. (Try, for example, to iterate the difference equation $x_{t+1} = ax_t(1 - x_t)$! Its behaviour is simple for small numerical values on a, but for larger values on a, consecutive x-values fluctuate chaotically.) In situations where this is not relevant, one may simplify the analysis by either imposing limits on the parameters or using the following approach.

It is often quite easy to transform difference equations into *differential equations*. These assume continuity, and their mathematical properties are simple (at least in one or two dimensions) and very well known. Thus, we will here transform the difference equation (1.1) into a differential equation. In many other cases in population genetics – in the next chapter, for example – this is not explicitly done or needed. Just knowing that such transformations *can* be made helps to calm any qualms about potential erratic behaviour of the functions involved.

If p_t is subtracted from both sides of (1.1), then the left side becomes $p_{t+1} - p_t = \Delta p$, where Δp is the change in p over one generation. We get

$$\Delta p = p_{t+1} - p_t = p_t(1 - \mu_{12}) + (1 - p_t)\mu_{21} - p_t$$

$$= -p_t\mu_{12} + \mu_{21} - p_t\mu_{21},$$

which can be rearranged into

$$\Delta p = \mu_{21} - (\mu_{12} + \mu_{21})p_t.$$

Since the change in p in one generation, Δp, will be a small value (this is where our assumption about the mutation rates being small enters), we can approximate Δp with dp/dt, where t is a measure of time given in generation numbers. Thus, we have

$$\frac{dp_t}{dt} = \mu_{21} - (\mu_{12} + \mu_{21})p_t. \tag{1.4}$$

This is a differential equation of a standard type that is easy to solve. We do so, quite generally, in Background 1:2, and find that it has solution:

$$p_t = \frac{\mu_{21}}{(\mu_{12} + \mu_{21})} + \left(p_0 - \frac{\mu_{21}}{(\mu_{12} + \mu_{21})} \right) e^{-(\mu_{12}+\mu_{21})t}. \tag{1.5a}$$

Here, p_0 is the frequency of A_1 in the assumed starting generation $t = 0$. Using expression (1.3), this can be rewritten as:

$$p_t = p_{eq} + (p_0 - p_{eq})e^{-(\mu_{12}+\mu_{21})t}. \tag{1.5b}$$

We have now found an expression for how p approaches its equilibrium value. Since the last term, $e^{-(\mu_{12}+\mu_{21})t}$, goes towards 0 for increasing values on t, it shows that whatever value the process is started from between 0 and 1, the frequency of A_1 will always move towards the internal equilibrium value p_{eq}. This makes p_{eq} into a *globally stable equilibrium* for process (1.1) and turns the characterized genetic polymorphism into a *balanced polymorphism*.

Half-time to equilibrium

To get a feeling for the dynamics of the process we have described, two methods are natural to use.

The first is to plot function (1.5) for specific values of μ_{12}, μ_{21} and p_0 – as we have done in Figure 1.1 – in order to visually follow the change in p over time. It can be seen that equilibrium p_{eq} is approached from above as well as below, but that the process is very slow – note the scale on the horizontal axis!

The second method is to derive an expression for the 'half-time' of the process (*i.e.* the time in generations that it takes to go half-way from any initial p_0 towards p_{eq}, a position we will designate by $p_{t_{1/2}}$). This position is reached when

$$p_{t_{1/2}} = p_0 + \frac{1}{2}(p_{eq} - p_0) = \frac{1}{2}(p_{eq} + p_0).$$

[11]

Derivation 1

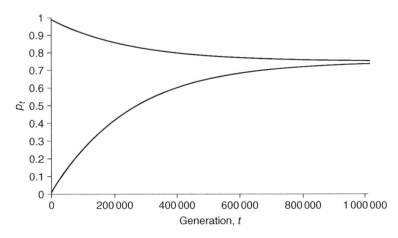

Figure 1.1 Graph showing how the frequency of allele A_1 changes with time as a function of mutations to and from the alternative allelic state A_2; see recursion system (1.1). The mutation rate from A_1 to A_2 is assumed to be $0.8 \cdot 10^{-5}$, while the rate of back-mutation is $0.2 \cdot 10^{-5}$. Two trajectories are shown: one for which evolution starts with an initial frequency for A_1 of 0.01, another for which the initial frequency of this allele is 0.99. It is seen that the equilibrium value, 0.8, is approached from above as well as below, but very slowly.

If this value for p_t is inserted into (1.5b), we get:

$$\frac{1}{2}(p_{eq} + p_0) = p_{eq} + (p_0 - p_{eq})e^{-(\mu_{12}+\mu_{21})t_{1/2}}$$

$$-p_{eq} + p_0 = -2(p_{eq} - p_0)e^{-(\mu_{12}+\mu_{21})t_{1/2}}$$

$$(p_{eq} - p_0) = 2(p_{eq} - p_0)e^{-(\mu_{12}+\mu_{21})t_{1/2}}$$

$$e^{-(\mu_{12}+\mu_{21})t_{1/2}} = \frac{1}{2}$$

$$-(\mu_{12} + \mu_{21})\, t_{1/2} = \ln\left(\frac{1}{2}\right)$$

$$-(\mu_{12} + \mu_{21})\, t_{1/2} = \ln(1) - \ln(2) = 0 - \ln(2) = -\ln(2)$$

$$t_{1/2} = \ln(2)/(\mu_{12} + \mu_{21}). \tag{1.6}$$

Since the mutation rates are small, this value, $t_{1/2}$, is guaranteed to be large – often, very large.

[12]

Summing up

In a situation where there are two alleles that mutate into each other at low rates, a large population in which genetic drift can be ignored will move towards a state where both alleles are present. The frequencies of the alleles at this equilibrium are given by a simple formula based on the mutation frequencies alone. From whatever state the population is started, this is the value towards it will evolve. The transition is described by an exponential function. A more detailed analysis shows that the speed with which the allele frequencies move towards their equilibrium values will in most situations be very slow. This means that the evolutionary force pushing the allele frequencies towards their equilibrium states is weak, and that other factors – such as genetic drift – may easily come to affect the process.

Interpretations, extensions and comments

The machinery of population genetics

With this derivation, the reader has become acquainted with the standard way of doing theoretical population genetics. The normal procedure is to consider an abstract situation – a model – that is similar to the empirically known world, but so well defined by its assumptions and so free from irrelevant factors that its dynamics can be captured by a simple equation (or a set thereof). The equation is then mathematically investigated until, hopefully, a concise result is reached that leads to a deeper and more satisfying feeling for how the model – and thereby the external world – functions. This mode of doing science is in principle the same as used by Galileo when, around the year 1600, he analysed the dynamics of physical phenomena such as falling balls and swinging pendulums. Since then, this has become the basic method of doing formal theory in the natural sciences.

The productive use of serious mathematical thinking in population genetics was initiated by Ronald A. Fisher in 1922. Soon thereafter, J. B. S. Haldane, starting in 1924, published an extensive series of articles on the effects of selection under different modes of inheritance. The third of the 'founding fathers of mathematical population genetics', Sewall Wright, entered the field with a broad synthesizing paper in 1931. In the present book, we consider many of the questions raised by Fisher in his wide-ranging publications, while our methodology is closer to the modelling approach used by Haldane and Wright. There were, of course, other scientists who contributed to the formation of population genetics as a well-defined theoretical subject, but it is from the scientific creativity of Fisher, Haldane and Wright that the foundation of the field derives. The early history of theoretical population genetics is interestingly described by Provine (1971) and Edwards (2000).

To practise theoretical population genetics, one must attain a feeling for which abstractions it is reasonable to make, how the equations are best formulated and solved, and – finally, but most importantly – what interpretations to draw from the results with respect to real biological situations. The focus in this book is on the formulation and mathematical analysis of specific questions, but we try to be clear also about which assumptions are made when nature is abstracted into our models.

Nucleotide mutations

The situation we have considered in this chapter is very simple – mutations back and forth between two allelic states unaffected by selection – and the result obtained is the expected one: when genetic drift can be ignored, the population will evolve towards a stable polymorphic equilibrium state. The relevance of such a very simple model may seem limited. However, only a small extension of the starting assumptions is needed to change the situation into a biologically highly relevant one. Assume instead mutations to and from *four* different allelic states! Let us call these allelic states– to no one's surprise – *A, C, T* and *G,* and we suddenly have a model for the evolution of nucleotides in a large population. The assumption that all genotypes have the same fitness (*i.e.* that the genetic variation is neutral), we retain from before.

It is easy to believe and prove (Crow and Kimura, 1970) that under these extended assumptions, the model population will again evolve towards an internal equilibrium at which the four nucleotides are present at well-defined proportions determined by the mutation rates between them (their number is twelve, since each nucleotide may mutate to any of the others). The equilibrium proportions can, for example, be used to calculate the expected *AT/GC* ratio at neutral sites.

[14]

Actually, it is biologically even more relevant to base the analysis on the sixteen possible nucleotide *pairs* and the role of mutations between them. This is because it is well known that the mutation behaviour of a nucleotide is strongly influenced by its nearest neighbour. The pair $3'-GC-5'$, for example, mutates in a different way and with different rates compared to the similar pair $3'-CG-5'$. Again, an internal equilibrium of frequencies is expected, which can easily be calculated if all the rates for the different mutation transitions are known. Such calculations are commonly made in computer programs used to analyse phylogenetic data.

Neutral evolution and an alternative method of analysis

Let us return to the results we obtained for the dynamics of two alleles.

Mutation frequencies are normally very small, usually of the order of 10^{-6} per locus and generation if by 'locus' we roughly mean a transcribed and translated gene, and maybe two or three orders of magnitude lower if we consider the locus to be a single nucleotide. Because of this, the expected rate of change in allele frequencies will be very slow and will have very long half-times to restore any deviations from the expected equilibrium. The equilibrium result given by expression (1.3) is, in textbooks of genetics, often used to introduce the idea of a stable equilibrium. In reality, however, chance effects will often overrule the weak mutation force and thereby influence the allele frequencies. The mathematically expected equilibrium frequency has therefore often only a weak predictive value when applied to single loci. Indeed, in many real situations, a locus of the considered kind will contain only a single allele – one of the alleles is then *fixed*, according to standard parlance.

This fact is, however, no reason for us to question the value of the population genetic analysis that we have performed in this chapter. The abstract version of reality – the model – that we formulated can be approached and analysed in other ways than the one we presented. Let us here sketch, with a new analytical procedure, an argument which shows the great and very general importance of expression (1.3). The argument is based on the fact that the expected number of throws needed to obtain a specific value, say a 4, with a fair die is six (*i.e.* the reverse of the probability to get a 4 in any of the throws), and that, in general, the expected number of times something must be repeated to produce an outcome with intrinsic probability p is exactly the inverse of p; that is, $1/p$. (This result is further described and developed in Background 4:2.)

Consider a gene copy of allelic type A_1. The gene copy is a replica of a gene copy in the earlier generation. That gene copy was probably of allelic type A_1 too, but we cannot be altogether sure since a mutation from A_2 may have occurred during the transmission between generations; the probability of this, according to our basic assumptions, is μ_{21}. The further back we go in generations, the larger the chance that the ancestral gene copy was of type A_2. Can we say anything about how many generations we on average must go back for this shift in allelic type to occur? Well, according to the logic just outlined for the perfect die, the expected number of generations back in time equals the inverse of the relevant mutation rate; that is, μ_{21}^{-1}. Similarly, a gene copy of current allelic type A_2 will have been transmitted in this state for, on average, μ_{12}^{-1} generations.

This implies that if we follow a gene copy through its transmission between very many generations, then its allelic state will have been A_1 for an average of μ_{12}^{-1} generations and A_2 for an average of μ_{21}^{-1} generations. If we think of ourselves as testing the process at a random moment in time, then it is clear that the probability that we will find an A_1 allele at the studied site must be $\mu_{12}^{-1}/(\mu_{12}^{-1} + \mu_{21}^{-1})$. Multiplying 'above' (the numerator) and 'below' (the denominator) with $\mu_{12}\mu_{21}$ produces the value $\mu_{21}/(\mu_{12} + \mu_{21})$, identical to the equilibrium value (1.3).

Thus, the equilibrium frequency (1.3) emerges as an important result also in this alternative analysis of our basic model. The importance of the alternative derivation is substantial, since it demonstrates that at any locus following the mutation process assumed, the probability of finding an allele of type A_1 is the same as the expected equilibrium frequency of this allele in a very large population. And this holds even for limited populations, where there is a chance that the population at the time of investigation in fact consists of only this one allele, or the other. Therefore, even if genetic drift will cause the locus to deviate from the expected equilibrium frequencies, the probabilities of finding the different alleles at the locus do not change.

This detour makes us understand how rich the possibilities of population genetics are. A specific model can be used to produce important results, but also to indicate their limits; we demonstrated this when we first calculated the equilibrium value and then indicated the weak strength of attraction to this equilibrium. In addition, it can be analysed with alternative methods to produce results that support and reinforce one another, as when we showed that the equilibrium frequency turns out also to be the expected value for this allelic state under genetic drift.

Background 1:1 Greek letters

Some Greek letters commonly used in population genetics and mathematics:

Letter	Pronounced
α	alpha
β	beta
δ, Δ	delta
θ	theta
λ	lambda
μ	mu
π	pi
σ	sigma
$\emptyset$	phi

Background 1:2 A common differential equation and its solution

The material presented here is very useful and is referred to in many of the coming derivations. The presentation should be easy to follow for anyone who has taken an elementary course in calculus.

Recall that the simple differential equation $\frac{dy}{dx} = y$, where y is function of x, has as its solution all equations $y = Ce^x$, where C is any real number. That this is the case can be demonstrated by differentiating the suggested solution Ce^x, which gives $\frac{d(Ce^x)}{dx} = Ce^x$, and, thus, $\frac{dy}{dx} = y$.

Remember also the following simple differentiation result:

$$\frac{d}{dx}e^{ax+b} = e^{ax+b} \cdot \frac{d}{dx}(ax+b) = ae^{ax+b}.$$

From these two formulae can be seen that the commonly occurring differential equation

$$\frac{dy}{dx} = b - ay, \text{ where } a, b > 0, \tag{1}$$

has as its solution all equations

$$y(x) = \frac{b}{a} + Ce^{-ax}, \tag{2}$$

where C is any real number.

Equation (1) is much used because it describes a process where the change in y with respect to x is a positive constant minus a term which directly depends on the size of y. Many important natural processes can – at least to the first approximation – be modelled with this equation. The classic equation for *logistic growth* of a bacterial population in a closed test tube, for example, follows from assuming that the growth rate of the bacteria decreases linearly with their number and concentration in the tube.

To ascertain that all equations of the form given by (2) are solutions to (1), one needs only find their derivative, $\frac{d}{dx}\left(\frac{b}{a} + Ce^{-ax}\right) = -aCe^{-ax}$, and then insert it into the original differential equation. This leads to $-aCe^{-ax} = b - a\left(\frac{b}{a} + Ce^{-ax}\right)$, which simplifies to $-aCe^{-ax} = b - b - aCe^{-ax} = -aCe^{-ax}$.

Thus, all equations given by (2) are indeed solutions to the differential equation (1). Often, one is interested in finding the solution to (1) that has a known specific starting point, say y_0, when x equals 0 (x is then often identified with the flow of time, assumed to start at 0). The equation that fulfils this condition must satisfy $y_0 = \frac{b}{a} + Ce^{-a0} = \frac{b}{a} + C$. From this follows that C in this case must be equal to $y_0 - \frac{b}{a}$. The solution to equation (1) with *boundary condition* $y(0) = y_0$ is thus:

$$y = \frac{b}{a} + \left(y_0 - \frac{b}{a}\right)e^{-ax}. \tag{3}$$

This equation gives, as expected, the value y_0 for $x = 0$ and goes towards $\frac{b}{a}$ when $x \to \infty$.

Questions

All chapters based on derivations end with questions for the reader to ponder and answer. The authors' answers are given at the back of the book. Some of the questions have been chosen so that they not only exemplify and illustrate the topics treated in the chapter, but also extend the results obtained.

1. Assume that $\mu_{12} = 1 \cdot 10^{-6}$ and $\mu_{21} = 3 \cdot 10^{-6}$. What is the equilibrium frequency of A_1, p_{eq}? What if μ_{12} remains the same but $\mu_{21} = 3 \cdot 10^{-8}$?

2. Return to the first mutation rates assumed in Question 1. If the population starts at $p_0 = 0.50$, how many generations will it take to reach $p = 0.625$ in a very large population? (This question is quite easy, so think before you start making any major calculations.)

3. Use expression (1.5b) to derive a function of the form $t = f(p_t, p_0, \mu_{12}, \mu_{21})$ that gives the number in generations it takes for a population to evolve to the allele frequency p_t from a starting point p_0, given the other parameters (for simplicity's sake, p_{eq} may also be used in the formula, since it is in itself a simple expression of the mutation rates). Use the same mutation rates as in Question 1 to calculate how many generations it takes to go from $p_0 = 0.50$ to $p_t = 0.70$.

4. Go back and reconsider the assumptions used in the derivation. How does expression (1.1) change if we assume that there are no back-mutations; that is, if $\mu_{21} = 0$? (This would be a reasonable assumption if, for example, one studied deletion mutations of more than ten base pairs covering a particular gene. It would also illustrate the possibility that one of the assumed allelic states, in this case A_2, in fact consists of a large number of different, but nonspecified, alternatives.) What is the only stable equilibrium frequency for p?

5. Let $p(A)_t$ be the probability that the nucleotide base at a specific position is A ($=$ adenine) in generation t, and similarly for $p(G)_t, p(C)_t$ and $p(T)_t$. Assume further that the DNA bases mutate according to what is called 'the Kimura 2-parameter model' (after Kimura, 1980). In this model, the probability per generation that a transition occurs is α and

the probability that a transversion occurs is β (transitions are mutations between A and G or between C and T, while transversions are the other mutational steps; see Figure 1.2). Express $p(A)_t$ as a function of $p(A)_{t-1}, p(G)_{t-1}, p(C)_{t-1}$ and $p(T)_{t-1}$. After looking at Figure 1.2, make a guess at the equilibrium value for $p(A)_t$ when $t \to \infty$.

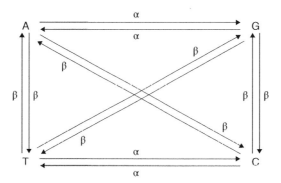

Figure 1.2 Illustration of the mutation relationships assumed in Question 5.

Derivation 2

Heterozygote advantage

Selection is the next systematic force to consider, after mutation. The first step in teaching selection is normally to consider a new allele that is better in every aspect than the allele that currently is the standard in the population. Here, we find this not sufficiently interesting, since the new allele is expected to increase in frequency until all gene copies in the population are of its particular kind – the allele will '*go to fixation*', as it is generally called. (The possibility that, instead, such a new beneficial allele will be lost to the population due to random genetic drift is treated in detail in Derivation 6.)

Instead, we will here assume that two different alleles produce the most fit genotype when they are combined in the same diploid individual (*i e.* in heterozygotes). Just like the situation in the preceding chapter with mutation and back-mutation, we expect a population with heterozygote advantage to move towards an equilibrium situation. The resulting evolutionary process cannot, however, be described as easily and fully as in the earlier chapter. We therefore introduce the standard method to investigate the stability of equilibria by analysing the effect of small perturbations.

The reader will from now on find abundant use in the book of the simple argument which says that if a term x is small then its squared term, x^2, is *very* small and can often be ignored relative to terms of size x. Or, to illustrate with a genetic example: If the mutation rate is 10^{-6} for a gene, then the probability that a gamete carries a newly mutated allele is 10^{-6}, while the probability that two randomly drawn gametes both carry new mutations is $(10^{-6})^2 = 10^{-12}$ – which is so small that in

Understanding Population Genetics, First Edition. Torbjörn Säll and Bengt O. Bengtsson.
© 2017 John Wiley & Sons Ltd. Published 2017 by John Wiley & Sons Ltd.

most situations it can be set to zero. Useful variations on this theme are presented in Background 2:2.

The chapter ends with a discussion of some of the hidden background assumptions and conventions normally employed in population genetics models.

Analysis

Assumptions

Consider a very large, random-mating population. Let there be a locus with two alleles A_1 and A_2. The relative fitnesses of the three genotypes A_1A_1, A_1A_2 and A_2A_2 are $1 - s$, 1 and $1 - t$, where we assume that s and t are greater than 0 but smaller than 1. These selection coefficients are here treated as fixed values; in Derivation 8 we will look more carefully at what can happen when the fitness of a genotype depends on the frequencies of the other genotypes in the population. Let the gametes that start a new generation have frequencies p and q, depending on whether they carry A_1 or A_2. Since these are the only possible gamete types, we have that $p + q = 1$.

Formulating the recursion system

Three diploid genotypes are possible given the assumptions, A_1A_1, A_1A_2 and A_2A_2, and at the start of the new generation their frequencies are in Hardy–Weinberg proportions p^2, $2pq$ and q^2 (the reader is reminded of this simple result, plus the meaning of 'random mating', in Background 2:1).

After selection, the frequencies of the three genotypes will be proportional to $(1 - s)p^2 : 2pq : (1 - t)q^2$. The sum of these values we call W, which we can express as follows:

$$W = (1 - s)p^2 + 2pq + (1 - t)q^2 = p^2 + 2pq + q^2 - sp^2 - tq^2$$
$$= 1 - sp^2 - tq^2.$$

The genotype frequencies after selection are, thus, $(1 - s)p^2/W$, $2pq/W$ and $(1 - t)q^2/W$. Their sum is exactly 1, which follows from the way we have defined W. These values are, indeed, true frequencies, and the W

expression functions here as what is called a 'normalizing factor'. When one follows the evolution of allele and genotype frequencies through an evolutionary process involving selection, a procedure similar to the one just used is often called for.

It is interesting to note that W in the present situation equals the mean fitness in the population in the formal sense of its *expected value*, which can be seen as follows:

$$Mean\ fitness = p^2 \cdot (1 - s) + 2pq \cdot 1 + q^2 \cdot (1 - t)$$
$$= p^2 + 2pq + q^2 - sp^2 - tq^2 = 1 - sp^2 - tq^2 = W.$$

(More on means and expectations can be found in Background 3:1.)

In the gametes that these genotypes produce, the two alleles will have the following frequencies, where the prime sign denotes the next generation:

$$p' = 1 \cdot (1 - s)p^2/W + \tfrac{1}{2} \cdot 2pq/W$$
$$q' = \tfrac{1}{2} \cdot 2pq/W + 1 \cdot (1 - t)q^2/W.$$

Or, expressed in words: All of the gametes produced by individuals having genotype A_1A_1 will be A_1, as will half of the gametes produced by individuals with genotype A_1A_2. Similarly, all the gametes produced by individuals having genotype A_2A_2 will be A_2, as will half of the gametes produced by individuals with genotype A_1A_2. (Note that we let t have a different meaning from that in Derivation 1 here, and that we use a new notation to mark the transition between generations; such changes in notations and meanings are very common in population genetics – they are perfectly all right, as long as one is explicit and consistent within the local context.) Simple rewriting gives

$$p' = p[p - sp + q]/W = p(1 - sp)/W$$
$$q' = q[p + q - tq]/W = q(1 - tq)/W.$$

Since in every generation the two allele frequencies add up to 1, we only need to focus on one of them. The recursion system following from our assumptions is, thus, fully specified by the expression

$$p' = p(1 - sp)/[1 - sp^2 - tq^2],$$

which can be written as

$$p' = \frac{p(1 - sp)}{1 - sp^2 - t(1 - p)^2}. \tag{2.1}$$

This recursion equation tells us what the frequency of A_1 will be at the start of the next generation, p', as a function of what it was at the start of the present generation, p, given selection coefficients s and t. Again, it is assumed that the population size is so large that sampling effects can be ignored. Other versions of the same recursion are given in Background 2:3, where selection is described with alternative sets of parameters, more useful for other purposes.

Expression (2.1) is structurally similar to equation (1.1), which was analysed with a differential equation in the preceding chapter. However, the expression reached here is mathematically more complex, in that p' now is a ratio in p and not just a linear function in p. We will therefore use another approach, common in population genetics, to study the behaviour of (2.1).

Finding the equilibrium points to the system

When does evolution stop? For what value on p does it follow that $p' = p = p_{eq}$? To answer these questions, we use the same method as in Derivation 1 and make the relevant substitution in the recursion equation (2.1):

$$p_{eq} = \frac{p_{eq}(1 - sp_{eq})}{1 - sp_{eq}^2 - t(1 - p_{eq})^2}. \tag{2.2}$$

This is a third-degree equation in p_{eq}. Such equations can be tricky to solve, but here all of its three roots are easily found. The first one is obvious:

$$\text{Root 1: } p_{eq} = 0.$$

After eliminating this root, equation (2.2) becomes

$$1 - sp_{eq} = 1 - sp_{eq}^2 - t(1 - p_{eq})^2,$$

which simplifies to

$$sp_{eq}(1 - p_{eq}) - t(1 - p_{eq})^2 = 0.$$

[26]

Thus, the second root is:

$$\text{Root 2: } p_{eq} = 1.$$

Eliminating this root simplifies the equation to

$$sp_{eq} - t(1 - p_{eq}) = 0,$$

which we can write

$$p_{eq}(s + t) = t.$$

The final root has thereby been found:

$$\text{Root 3: } p_{eq} = t/(s + t).$$

Of these three roots, the first two describe what are called *boundary equilibria*, since they are at the limits for our variable under investigation (since p_{eq} is a frequency, its range is between 0 and 1, these values included), while the third describes an *internal equilibrium* to the recursion system (for s and t greater than zero but smaller than one, this root falls always between 0 and 1). The existence of this equilibrium and its exact value was first given by Fisher (1922).

At the boundary equilibria where p_{eq} equals 0 or 1, the mean fitness in the population is $1 - t$ or $1 - s$, respectively. At the internal equilibrium, the mean fitness (using the expression for W) is:

$$1 - s\left(\frac{t}{s+t}\right)^2 - t\left(1 - \frac{t}{s+t}\right)^2 = 1 - s\left(\frac{t}{s+t}\right)^2 - t\left(\frac{s}{s+t}\right)^2$$

$$= 1 - \frac{st^2 + s^2 t}{(s+t)^2} = 1 - \frac{st}{s+t}. \qquad (2.3)$$

Later in the chapter we will discuss a particular and interesting implication of this result further.

Stability

To determine the state(s) to which the model will evolve, we have so far found the points where $p' = p$; that is, the situations where the alleles have constant frequencies over time. The next step is to determine whether these different equilibrium points are locally stable or unstable.

Stability condition

To check for the properties of the equilibria to (2.1), we introduce a general method for investigating the local stability in recursion systems. This method assumes that the process starts at a frequency slightly perturbed away from one of the equilibrium points, and it then studies whether this perturbation increases or decreases with time. If the perturbation increases in absolute size, then the system moves away from the investigated equilibrium, which will be called unstable, while if the perturbation decreases, then the system will move back towards the equilibrium, which will be called stable. Stability analyses can be made more elegantly than they are here, with the aid of more advanced mathematical tools – see Background 10:1. There is, however, a real value in doing these analyses from first principles, since a deeper insight into how the dynamic system functions is thereby gained.

Thus, let p_{eq} be one of the three equilibria to system (2.1) and look at what happens when the recursion is started at allele frequency $p_{eq} + x$. x can be greater or smaller than zero, but it must of course always be such that $p_{eq} + x$ remains within the interval between 0 and 1. The perturbation x is assumed to be so small that its squared value can be ignored. We still use expression (2.1) to follow the behaviour of the system over time; to do so, we let p equal $p_{eq} + x$ and p' equal $p_{eq} + x'$. Our first aim is to express x' in terms of x (together with s and t and the equilibrium value p_{eq}). This involves some algebraic manipulation – tedious but not really difficult – where ample use is made of the approximation rules described in Background 2:2.

The analysis starts by studying the enumerator of the right side of expression (2.1):

$$(p_{eq} + x)[1 - s(p_{eq} + x)] = p_{eq}[1 - s(p_{eq} + x)] + x[1 - s(p_{eq} + x)]$$
$$\approx p_{eq}(1 - sp_{eq}) - p_{eq}sx + x(1 - sp_{eq})$$
$$= p_{eq}(1 - sp_{eq}) + (1 - 2sp_{eq})x.$$

This, we express as $a + bx$, where $a = p_{eq}(1 - sp_{eq})$ and $b = 1 - 2sp_{eq}$.
The denominator becomes:

$$1 - s(p_{eq} + x)^2 - t(1 - p_{eq} - x)^2$$

$$\approx 1 - sp_{eq}^2 - 2sp_{eq}x - t(1 - p_{eq})^2 + 2t(1 - p_{eq})x$$
$$= 1 - sp_{eq}^2 - t(1 - p_{eq})^2 - 2[sp_{eq} - t(1 - p_{eq})]x.$$

This, we chose to express as $c + dx$, where $c = 1 - sp_{eq}^2 - t(1 - p_{eq})^2$ and $d = -2[sp_{eq} - t(1 - p_{eq})]$. We can now use approximation rule (3) from Background 2:2 to get:

$$p_{eq} + x'$$

$$= \frac{a + bx}{c + dx} \approx \frac{a}{c} - \frac{ad - bc}{c^2}x = \frac{p_{eq}(1 - sp_{eq})}{1 - sp_{eq}^2 - t(1 - p_{eq})^2}$$

$$- \frac{\{p_{eq}(1 - sp_{eq})(-2)[sp_{eq} - t(1 - p_{eq})] - (1 - 2sp_{eq})[1 - sp_{eq}^2 - t(1 - p_{eq})^2]\}x}{[1 - sp_{eq}^2 - t(1 - p_{eq})^2]^2}$$

$$= \frac{p_{eq}(1 - sp_{eq})}{1 - sp_{eq}^2 - t(1 - p_{eq})^2}$$

$$+ \frac{\{(1 - 2sp_{eq})[1 - sp_{eq}^2 - t(1 - p_{eq})^2] + 2p_{eq}(1 - sp_{eq})[sp_{eq} - t(1 - p_{eq})]\}x}{[1 - sp_{eq}^2 - t(1 - p_{eq})^2]^2}.$$

In this long expression, p_{eq} is one of the three equilibria to the recursion process, and for all three of these it holds (see 2.2) that

$$p_{eq} = \frac{p_{eq}(1 - sp_{eq})}{1 - sp_{eq}^2 - t(1 - p_{eq})^2}.$$

Therefore, the long expression can be simplified to

$$x' = Sx, \tag{2.4a}$$

where

$$S = \frac{(1 - 2sp_{eq})[1 - sp_{eq}^2 - t(1 - p_{eq})^2] + 2p_{eq}(1 - sp_{eq})[sp_{eq} - t(1 - p_{eq})]}{[1 - sp_{eq}^2 - t(1 - p_{eq})^2]^2}.$$

$$\tag{2.4b}$$

We have now reached our first goal. By evaluating S for the equilibrium values p_{eq}, we will know whether they are unstable (which happens if $|S| > 1$) or stable (which occurs for $|S| < 1$).

Derivation 2

Equilibrium $p_{eq} = 0$

Making the substitution $p_{eq} = 0$ in expression (2.4) leads to

$$x' = \frac{1}{1-t}x,$$

and we have proven that this equilibrium is unstable, since from the beginning we have assumed that t is greater than zero but smaller than 1. With this assumption, x' will always be numerically greater than x, which implies that the system will always evolve away from the equilibrium value $p_{eq} = 0$ if perturbed.

Equilibrium $p_{eq} = 1$

With the substitution $p_{eq} = 1$, expression (2.4) becomes instead

$$x' = \frac{1}{1-s}x,$$

as could be expected from symmetry considerations. Again, this result indicates that the equilibrium is unstable. Based on the last two results, we can say that our system must have a *protected polymorphism*, since the two possible fixation states (boundary equilibria) are both unstable.

Internal equilibrium

The stability of the internal equilibrium $p_{eq} = t/(s+t)$ remains to be determined. Making this substitution, while using the simple fact that $1 - p_{eq} = s/(s+t)$, we find that the numerator of S becomes

$$\left[1 - \frac{2st}{s+t}\right]\left[1 - \frac{st^2}{(s+t)^2} - \frac{s^2t}{(s+t)^2}\right] + 0$$

$$= \left[\frac{s+t-2st}{s+t}\right]\left[\frac{(s+t)^2 - st^2 - s^2t}{(s+t)^2}\right]$$

$$= \left[\frac{s+t-2st}{s+t}\right]\left[\frac{(s+t)^2 - st(t+s)}{(s+t)^2}\right] = \frac{(s+t-2st)(s+t-st)}{(s+t)^2}.$$

Similarly, the denominator of S becomes

$$\left[1 - \frac{st^2}{(s+t)^2} - \frac{s^2t}{(s+t)^2}\right]^2 = \left[\frac{(s+t)^2 - st(t+s)}{(s+t)^2}\right]^2 = \left[\frac{s+t-st}{s+t}\right]^2.$$

Putting these together, we get — with some simplifications — that

$$S = \frac{(s + t - 2st)(s + t - st)(s + t)^2}{(s + t)^2(s + t - st)^2} = \frac{s + t - 2st}{s + t - st} = \frac{s + t - st}{s + t - st}$$
$$- \frac{st}{s + t - st} = 1 - \frac{st}{s + t - st}.$$

The last expression, $\frac{st}{s+t-st}$, is positive and always smaller than 1 when s and t are positive and smaller than 1. Thus, $|S| < 1$ and we have now proven that the internal equilibrium is always stable.

The value $||S| - 1|$, in this case $st/(s + t - st)$, is a measure of the strength of the stability of the equilibrium. The larger this value, the faster the system will move towards the equilibrium point after a perturbation.

From a mathematical point of view, we have now studied the *local* stability of the equilibrium points to our dynamic process. In the preceding analysis of the mutation–mutation balance, the more difficult question of *global* stability was approached and solved. In the present situation, a local analysis is perfectly satisfying, since we expect the system to evolve towards an equilibrium point, and in this case only one of them is locally stable. (A critical reader may ask: *Why* do we expect the system to evolve towards an equilibrium point? Can we be absolutely sure that the allele frequencies will not circle round or jump here and there in an erratic fashion? Our answer was hinted at in the preceding chapter: When the parameters s and t are suitably small, the difference equation can be approximated with a differential equation which will be well-behaved.)

Summing up

There are two alleles at a locus in a very large population. Mutations are ignored, and the heterozygotes are more fit than the two kinds of homozygotes. The system has three equilibrium points. Two are what mathematicians would call 'trivial boundary equilibria', where one or the other of the alleles is fixed. The third, internal, equilibrium point is given by the strengths of selection against the homozygotes. This equilibrium is locally stable, which implies that if the population is disturbed away from this state, it will evolve back towards it. The boundary equilibria are locally

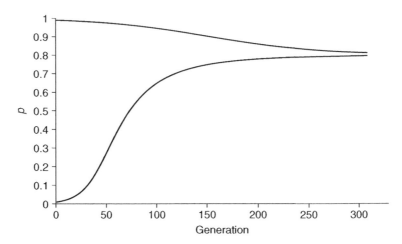

Figure 2.1 Change over time in the frequency of allele A_1 as a function of selection favouring heterozygotes (see recursion system (2.1)), assuming s to be 0.002 and t to be 0.008. The equilibrium value, 0.8, is approached – from above as well as from below – much more rapidly than in Figure 1.1.

unstable, which implies that small perturbations will cause evolution to move away from these states. The dynamics of the model is illustrated in Figure 2.1.

Interpretations, extensions and comments

Heterozygote advantage and disadvantage

Some examples of heterozygote advantage are well known, in particular the situation where mutant alleles increase the fitness of heterozygotes by providing improved resistance to malaria but decrease the fitness of mutant homozygotes due to various genetic illnesses (Hedrick, 2011b), but many other biological systems have also been suggested to follow this logic (see *e.g.* Mead *et al.*, 2008). More common – but actually much less written about – is, however, the opposite situation: heterozygote disadvantage.

This situation is common because all structural chromosome mutations, such as inversions and translocations, decrease in heterozygous form the fitness of their carriers due to the problems they cause in meiosis, even when they are perfectly genetically balanced (why this is so may be learnt from any genetic textbook). It is, in fact, possible to study the population genetics of such genetic variation using the preceding analysis almost unchanged. The only alteration needed is to assume that s and t are negative instead of positive values. This makes the fitness of the homozygotes greater than the fitness of the heterozygotes, and the model thus becomes a model for how the population evolves under heterozygote disadvantage. The same basic equations (2.1 and 2.2) can be used; the only thing one needs to remember is that now $0 > s, t$. The three equilibria $p_{eq} = 0$, $p_{eq} = t/(s + t)$ and $p_{eq} = 1$ are therefore the same as before. The important difference comes instead with the stability analyses. Now it is found that the two boundary equilibria are stable and the internal equilibrium is unstable, as you can easily check by yourself. (We are now also in a situation where one or more fitness value is greater than 1. This is perfectly all right, since we are only interested in the relative proportions between the different fitness values. More on alternative ways of describing fitness relationships can be found in Background 2:3.)

New chromosome mutations are thus normally selected against when rare, and if such a mutation happens to reach a certain frequency by random drift, for example, then its frequency is expected to decrease in future generations. Why chromosome mutations nevertheless sometimes increase to fixation and thereby change the population's karyotype we will not discuss here. Let us instead contemplate what happens when a new chromosome mutation occurs that is deleterious in heterozygote form but better, in one fitness component or other, than the standard when homozygous (thus, $0 > s > t$). From an external perspective, it would be 'best for the population' if this mutation were to spread until it reached fixation, since the fitness in the population would then increase from $1 + |s|$ to $1 + |t|$ − but this will not normally happen. The new chromosome mutation will always be selected against as long as it is rare, and it is therefore not expected to increase in frequency.

By this example, we are reminded that evolution by natural selection is not a machine that automatically leads to an absolute optimum. In real life, the Darwinian idea must always function under Mendelian logic, and both of these processes act in present time and are blind to long-term consequences.

[33]

Increase in mean fitness

We expect, however, that under the simple assumptions outlined in the beginning of the chapter, the mean fitness in the population will always increase from generation to generation.

The idea is in this case easily checked, at least in theory. The only thing needed is to compare the mean fitness of the new generation to the mean fitness in the preceding generation. By standard algebra (and quite a lot of work; interested readers can check themselves!), it is found that

$$W' - W = 2p(1-p)\left[\frac{t}{s+t} - p\right]^2 [2(1-t) - (s-3t)p - (s+t)p^2].$$

All the factors in this expression are strictly positive for valid values on p outside the equilibrium points, irrespective of whether s and t are positive or negative; at the equilibrium points, evolution stops and the population mean fitness stays the same generation after generation. This result shows that the mean fitness in the population never decreases when selection acts on two alleles in a population and the fitness values are fixed.

It is possible to go one step further and assume that there are many alleles at the locus and that every possible genotype has its own fixed fitness; it is then still the case that the population mean fitness will never decrease between generations. This perfectly reasonable and expected result is, however, not simple to derive. It was first proposed by Mandel and Hughes (1958), and later proven by a number of authors; the mathematically most satisfying proof (though technically advanced) was given by Kingman (1961).

Origin of the neutral theory of molecular evolution

The dynamics that derives from there being heterozygote advantage at a locus in a random mating population attracted attention and interest very early in the history of population genetics, and the theoretical results given in this chapter can be found in the early writings of all the founding fathers of population genetics (Fisher, 1922; Haldane, 1924; Wright, 1931). This culminated in a situation where it was more or less taken for granted that if a polymorphism existed in a population, then it was due to some kind of heterozygote advantage. One must remember that for a long time, very few genetic polymorphisms – mainly blood-groups – were known in species such as humans.

The situation started to change after the Second World War, with the introduction of biochemical separation techniques such as enzyme electrophoresis. During the 1960s, more and more data became available which showed that many genes in humans and in Drosophila are represented by more than one allele. These results were of great interest, in particular since with more variability available for analysis, better and more informative genetic studies could be performed. But they were also problematic, since – as we have seen – a population with heterozygote advantage at a locus with two alleles has its mean fitness reduced by $\frac{st}{s+t}$ compared to the highest possible value, 1 (see expression (2.3)). This reduction, generally called *the segregation load*, would normally not matter much. But if such a load were caused by every single polymorphism, and there were thousands of polymorphisms in the species, then the overall effect on the population would be staggering. Thus, Lewontin and Hubby (1966) ended their path-breaking description of extensive isozyme variation in *Drosophila pseudoobscura* with the following calculation: If s and t were about 0.10, and the number of loci under selection in the species is 2000, then mean relative fitness would be $\left(1 - \frac{0.1^2}{0.1+0.1}\right)^{2000} \approx 10^{-46}$, an impossibly low number.

How populations might cope with high amounts of segregation loads due to heterozygote advantage became one of the great questions in population genetics around 1970. Out of heated discussions grew the idea, which today imbues almost all thinking and methods in bioinformatics, that most of the molecular variation that segregates in populations is *not* in a balanced state due to heterozygote advantage but is selectively neutral, or nearly so. With this explanation, forcefully promoted by Motoo Kimura (1968, and in extensive form 1983), the problem with the load disappears and the amount of segregating alleles in a population is under no selective limit.

Here we see how the theoretical results that we derived earlier were once scientifically helpful in a slightly paradoxical way. They showed not only that certain baffling phenomena, such as sickle cell anaemia, can be understood, but also that if this type of explanation is stretched too far, then it produces results that are severely problematic and in need of a reinterpretation. One way by which theoretical models can inform us about the world is, thus, by forcing us to reconsider our standard way of explaining it. This is the point made by the American philosopher of science Thomas Kuhn in his discussion on the use of thought experiments in physics (see Kuhn, 1977).

Underlying assumptions in population genetics models

Now, when we have come this far in the book, it may be the right time to comment on some of the underlying structures that we take for granted in our analyses and that the reader might wonder about. These structures have to do with the way we build and describe the models that are analysed mathematically.

In this chapter, we focus on organisms with different fitness values, and these organisms we assume to be diploid (though we also take for granted that in the species there is a haploid phase as part of the reproductive cycle). In the preceding chapter, on the other hand, we concentrated on the genetic material itself and paid no particular attention to the ploidy level of the carriers. The organisms could therefore be haploid, but they could also be diploid or polyploid (the mutation models discussed there may, thus, be relevant for prokaryotes as well as for all kinds of eukaryotes). In the forthcoming derivations, we will not spend much time on describing exactly which organisms the assumed models may be relevant for – any interested reader will understand what types of organisms may fit the model assumptions.

Similarly, we will throughout the book assume that generations follow one another in a neat, orderly and discrete fashion. (Standard population genetic modelling is, thus, based on the 'Markov property', implying that the state at a particular moment in time of a studied process is determined by the state in the directly preceding moment.) As an alternative, one could instead assume that birth, reproduction and death occur all the time in a population, and express our population genetics questions within such a richer and more realistic framework. The results would, however, be very similar to the ones we derive here, while the mathematics would be more cumbersome; this explains why we limit our attention to distinct generations. Charlesworth (1994) gives a detailed introduction to these considerations.

When we discuss selection, we normally treat it as happening in the form of viability selection (*i.e.* individuals with certain genotypes survive better or less well than individuals with other genotypes). Selection acting through other fitness components is, thus, generally ignored (except in Derivations 8 and 10, where offspring production is more explicitly discussed). This limitation is, again, used for practical purposes. Bodmer (1965), for example, showed that one gets identical results if one assumes that individuals differ in fertility rather than in viability, as long as the relative output of matings is given by the product of the female's and the male's respective fertilities. The concentration on individual viability also implies that such interesting questions in sexual selection as mate-choice and within-sex competition are ignored. They often lead to frequency-dependent selective interactions of the kind discussed in Derivation 8.

Finally, it should be noted that it normally does not matter in which order events are assumed to happen within the reproductive cycle. Thus, in some later chapters it is assumed that mutation occurs only during the gametic, haploid, phase of the meiotic cycle, while selection occurs only during the zygotic, diploid, phase – these restricting assumptions make the analysis simpler but do not significantly affect the results.

Background 2:1 Reminder: Hardy–Weinberg proportions

In 1908, the English mathematician G. H. Hardy and the German physician Wilhelm Weinberg independently described the following insight, taught today in all elementary textbooks on genetics: Let there be two kinds of gametes, those carrying allele A_1 of a gene and those carrying the alternative allele A_2, and let these occur in frequencies p and q, where $p + q = 1$. If the gametes form diploids by uniting at random two and two, then the three types of possible diploid genotypes, A_1A_1, A_1A_2 and A_2A_2, will be formed in relative proportions p^2, $2pq$ and q^2. These three values add, as expected, to 1, since $p^2 + 2pq + q^2 = (p + q)^2 = 1^2 = 1$.

The random union of gametes is often described as occurring via '*random mating*'; that is, mating between pairs of diploids coming together independently of the considered genetic variation.

The structure of these proportions makes us immediately realize that homozygotes for rare alleles will be *very* rare (if q is small, then q^2 will be very small). This implies that, under random mating, rare alleles occur almost exclusively in heterozygotes and not in homozygotes.

If there are many allelic types of the considered gene, say $A_1, A_2, \ldots A_k$, with frequencies $p_1, p_2, \ldots p_k$, then under random mating the frequency of homozygotes A_iA_i will be p_i^2 and that of heterozygotes A_iA_j will be $2p_ip_j$.

It is quite common in scientific texts today to abbreviate 'Hardy–Weinberg' to 'HW'.

The principle used to derive the Hardy–Weinberg proportions can be used also when pairwise combinations of other kinds of objects are considered. Thus, if there are two kinds of animals in a population with frequencies x and y (identical in the two sexes), then the frequency with which mixed mating pairs are formed will be $2xy$, given that encounters occur at random.

Finally, we recall the standard result, that if there is *inbreeding* in the population due to some matings between close relatives, then the proportion of genotypes in the two-allele case will be $p^2 + Fpq$, $2pq − 2Fpq$ and $q^2 + Fpq$. Here, F denotes the average of the individuals' inbreeding coefficients.

Background 2:2 Some useful approximations

We first consider how mathematical expressions can be simplified by ignoring very small numbers. We then describe some approximations of the exponential function.

If x is a small number close to zero, then its square, x^2, is *very* small. This is even more correct for higher powers of x. This implies that if we have found that the effect of x on something is given by expression $f(x) = a + bx + cx^2 + dx^3 + \ldots$ (where $a, b, \ldots$ do not grow large as x approaches zero and b is different from zero), then to a first-order approximation we can ignore all higher terms in x – as long as x stays small – and write:

$$f(x) \approx a + bx. \tag{1}$$

Using the same logic, two more complicated but useful approximation rules, valid for small values on x, can be derived:

$$\frac{ax}{b + cx} = \frac{ax/b}{1 + cx/b} = \frac{ax/b(1 + cx/b)}{1 + cx/b} - \frac{acx^2/b^2}{1 + cx/b} = \frac{ax}{b} - x^2 \left[\frac{ac/b^2}{1 + cx/b} \right] \approx \frac{a}{b}x, \tag{2}$$

and:

$$\frac{a + bx}{c + dx} = \frac{a}{c + dx} + \frac{bx}{c + dx} \approx \frac{a}{c + dx} + \frac{bx}{c}$$

$$= \frac{(ac/c) + (adx/c) - (adx/c)}{c + dx} + \frac{bx}{c}$$

$$= \frac{(a/c)(c + dx)}{c + dx} - \frac{(adx/c)}{c + dx} + \frac{bx}{c} = \frac{a}{c} - \frac{adx}{c(c + dx)} + \frac{bx}{c}$$

$$\approx \frac{a}{c} - \frac{adx}{c^2} + \frac{bx}{c} = \frac{a}{c} - \frac{ad - bc}{c^2}x. \tag{3}$$

In both cases, we reduce more complex relationships to simple linear functions around the point $x = 0$. In common parlance, the expressions have become linearized.

(A small technical note: The first time we use an approximation, we mark its occurrence with the $\approx$ sign, but later we return back to the = sign, just as we have done here.)

Let us now give some approximations relevant for the exponential function. Remember that

$$e = \lim_{n \to \infty} \left(1 + \frac{1}{n}\right)^n,$$

and that for fixed values on x

$$\lim_{n \to \infty} \left(1 + \frac{x}{n}\right)^n = e^x.$$

Here, x may be positive or negative.

Without special proof, we then accept the following two related approximations, which hold for small positive or negative values of x. The first is that:

$$e^x \approx 1 + x. \tag{4}$$

The second is that:

$$(1 + x)^t \approx e^{xt}. \tag{5}$$

Background 2:3 Modelling selection

Selection on genotypes can be envisioned and modelled in many different ways. Here, we describe some alternative approaches to generalizing the method used so far, but also to making later derivations easier.

Since we are only interested in relative fitness and there are three genotypes to consider (assuming two alleles in a diploid organism), only two parameters are needed for a full description of selection. In the present context, where heterozygote advantage is discussed, the two positive parameters s and t have been used to describe the decrease in fitness for the two types of homozygotes. But we have also seen that heterozygote disadvantage can be treated by the same set-up, if only s and t are given negative values.

In Derivation 6, we will study the probability of fixation of a mutation, A_1, that is favoured by selection but that may be lost from the population due to random genetic drift. The way positive selection is handled there can be written as follows:

Genotype	A_1A_1	A_1A_2	A_2A_2
Fitness	1	$1 - s/2$	$1 - s$

To be a model of selection favouring allele A_1, it is posited that $s > 0$. This is, in fact, a more constrained model of selection than the one we gave earlier, since it is assumed that the fitness of the heterozygotes is exactly intermediary to the fitness of the two types of homozygotes; a single parameter is, thus, sufficient to describe the process of selection. Using the language of quantitative genetics, one would here talk about the allele A_1 as having an *additive effect* on fitness. We will now consider what recurrence equation this parameterization gives rise to.

Let p be the relative frequency of the A_1 allele and let $q = 1 - p$. If there is random mating, then the mean fitness in the population will be $W = p^2 \cdot 1 + 2pq \cdot (1 - s/2) + q^2 \cdot (1 - s) = 1 - spq - sq^2$. The allele frequency in the next generation can, thus, be written:

$$p' = \frac{p^2 + \frac{1}{2} \cdot 2pq(1 - s/2)}{1 - spq - sq^2} = \frac{p(p + q - sq/2)}{1 - sq(p + q)} = \frac{p(1 - sq/2)}{1 - sq}. \qquad (1)$$

This is, in effect, a simplified version of recursion system (2.1), based on a different way of specifying the fitness values. With some maths, this can be seen as follows:

Substitute in the model of heterozygote advantage s by $-\frac{s}{2-s}$ and t by $\frac{s}{2-s}$, and assume s to be positive. The fitness values for the three genotypes then become $1 + \frac{s}{2-s}$, 1 and $1 - \frac{s}{2-s}$, which – since we are only interested in relative values – correspond to the fitness scheme 1, $1 - s/2$, $1 - s$. If the same substitution is made in expression (2.1), the result looks a bit messy at first, but after simplification, and remembering that $q = 1 - p$, expression (1) appears. This shows that the parametrization used while analysing heterozygote advantage is capable of capturing the effects of *all* kinds of selection, if only the parameters s and t are allowed to take negative as well as positive values.

A very common way of modelling selection at a locus is highly similar to what we just have presented but slightly more general:

Genotype	A_1A_1	A_1A_2	A_2A_2
Fitness	1	$1 - hs$	$1 - s$

Derivation 2

This way of formulating selection leads to the following difference equation:

$$p' = \frac{p^2 + pq - hspq}{1 - 2hspq - sq^2} = \frac{p(p + q - hsq)}{1 - 2hspq - sq^2} = \frac{p(1 - hsq)}{1 - 2hspq - sq^2}. \tag{2}$$

New here, compared to the earlier case, is that the fitness of the heterozygotes is no longer strictly intermediary to the homozygotes but is under the influence of a potential dominance relationship between the alleles. When selection is formulated like this, s is normally called the selection coefficient and h the degree of dominance between the alleles. For $h = 0$ or 1, there is strict dominance at the locus, with one of the alleles being dominant (with respect to the fitness trait) relative to the other, recessive, allele. This way of describing selection at a locus becomes useful when, for example, the balance between mutation and selection is to be studied. The set-up is also used in Derivation 9, where we consider how selection on a quantitative trait affects the alleles at a contributing locus.

Questions

1. Why is there such a difference in the speed with which the equilibrium is approached between the figures in this and in the preceding chapter? And why are there boundary equilibria to the evolutionary system in this chapter but not in the preceding chapter?

2. Regard the population mean fitness W in the case of heterozygote advantage as a function of p (use the expression $W = 1 - sp^2 - tq^2$, where $q = 1 - p$). Describe the shape of this function over $0 \le p \le 1$. In particular, where does it have its maximum?

3. Since fitness values are relative, the evolution of a positive allele with additive effect might as well be modelled by fitness scheme $1, 1 - s$, $1 - 2s$ as by fitness scheme $1 + 2s, 1 + s, 1$ (which in its turn differs from the similar example analysed in Background 2:3). Find the recursion equations for the selected allele in these two cases, and show that they are identical when s is small.

4. In Background 2:3, it is claimed that result (1) corresponds to expression (2.1), on which the main derivation is based. Complete the outline of the sketched proof with the relevant calculations.

5. Let p be the frequency of A_1 in a random-mating population with heterozygote advantage of the kind analysed in our main derivation, and let W be the population mean fitness as in Question 2. Show that under these assumptions, $\Delta p = p' - p$ equals $Cov[X, Y]/2W$, where X is the number of A_1 copies that an individual carries and Y is its fitness. This implies that at the equilibrium where the frequency of the A_1 allele stays unchanged over time, there is no correlation between the number of alleles – 0, 1 or 2 – that an individual carries and the fitness of the individual. That the rate of evolution of an allele depends directly on the covariance between the copy number of this allele and the fitness of its carriers was shown for a wide range of models by Price (1970). (If you are uncertain about the concepts of covariance and correlation, you can read more about them in Background 7:1.)

Derivation 3

Breakdown of linkage disequilibrium

Given the size of genomes and the inevitability of mutations, any two individuals in a population will differ with respect to many genetic sites, loci. One of the tasks of population genetics is to find tools for succinct descriptions of such genetic variation.

To characterize the situation at a particular locus, the natural choice is to use the number and frequencies of the alleles at the locus. How to describe the associations and correlations that may exist *between* alleles at *different* loci is, however, not equally obvious. Here, many possibilities can be envisaged. In this chapter, we introduce the standard measure of linkage disequilibrium, D, and derive a key result concerning its rate of breakdown due to recombination in the absence of other evolutionary factors. The result explains why D, and measures closely related to D, have become such useful tools for measuring the association between alleles at distinct loci.

Analysis

Assumptions, notations and definitions

Consider a very large, randomly-mating population of an organism in which meiosis occurs as an obligatory part in the life cycle. Here, we are interested in the association between alleles in the haploid gametic phase of the cycle, and for this purpose we consider two loci, A with alleles A_1 and A_2 and B with alleles B_1 and B_2. The recombination frequency between the loci is r ($0 \leq r \leq 1/2$). (In human genetics, the Greek letter θ is often used for this purpose.) Thus, a gamete can have one of the

Understanding Population Genetics, First Edition. Torbjörn Säll and Bengt O. Bengtsson. © 2017 John Wiley & Sons Ltd. Published 2017 by John Wiley & Sons Ltd.

following four allelic combinations, also called *haplotypes*: A_1B_1, A_1B_2, A_2B_1 and A_2B_2. No mutations occur in the system, and there are likewise no fitness differences between individuals or gametes with different genotypes. Our aim is to find a suitable way to describe the relationship between the allele frequencies at the two loci and the frequencies of the different haplotypes, and then to follow how this relationship is changed over time by recombination.

We start by introducing a convenient notation for the relative frequencies of gamete types:

$$P_{11} = p(A_1B_1), \tag{3.1a}$$

$$P_{12} = p(A_1B_2), \tag{3.1b}$$

$$P_{21} = p(A_2B_1) \text{ and} \tag{3.1c}$$

$$P_{22} = p(A_2B_2). \tag{3.1d}$$

These frequencies will be called *haplotype frequencies*. The allele frequencies can be expressed in terms of these haplotype frequencies in the following way:

$$p(A_1) = P_{11} + P_{12}, \tag{3.2a}$$

$$p(A_2) = P_{21} + P_{22}, \tag{3.2b}$$

$$p(B_1) = P_{11} + P_{21} \text{ and} \tag{3.2c}$$

$$p(B_2) = P_{12} + P_{22}. \tag{3.2d}$$

With the notation $p(x)$, we here – and in many later instances in the book – denote 'the relative frequency of x'; the letter p is, of course, chosen to associate with the word 'probability'.

If the co-occurrence of alleles in haplotypes is independent (in the statistical sense), then *linkage equilibrium* (also called gametic equilibrium) is said to occur. Thus, in linkage equilibrium, the following relationships between haplotype and allele frequencies hold:

$$P_{11} = p(A_1)p(B_1), \tag{3.3a}$$

$$P_{12} = p(A_1)p(B_2), \tag{3.3b}$$

$$P_{21} = p(A_2)p(B_1) \text{ and} \qquad (3.3c)$$

$$P_{22} = p(A_2)p(B_2). \qquad (3.3d)$$

If, on the other hand, there is any deviation in the haplotype frequencies from expressions (3.3a–d), then A and B are said to be in *linkage disequilibrium* (often abbreviated LD).

We introduce the standard measure of linkage disequilibrium, D, by

$$P_{11} = p(A_1)p(B_1) + D, \qquad (3.4a)$$

where D may be positive or negative. It then follows from (3.2) that

$$P_{12} = p(A_1) - P_{11} = p(A_1) - p(A_1)p(B_1) - D$$
$$= p(A_1)[1 - p(B_1)] - D = p(A_1)p(B_2) - D, \qquad (3.4b)$$

and similarly that

$$P_{21} = p(B_1) - P_{11} = p(B_1) - p(A_1)p(B_1) - D$$
$$= p(B_1)[1 - p(A_1)] - D - p(A_2)p(B_1) - D \qquad (3.4c)$$

and

$$P_{22} = p(B_2) - P_{12} = p(B_2) - p(A_1)p(B_2) + D$$
$$= p(B_2)[1 - p(A_1)] + D = p(A_2)p(B_2) + D. \qquad (3.4d)$$

That a single parameter is sufficient to describe the associations between the alleles in the four types of gametes is due to the fact that, since there are four proportions that sum to unity, there are three degrees of freedom. The allele frequencies $p(A_1)$ and $p(B_1)$ represent one degree of freedom each; thus, there is only one degree of freedom left – this is the one represented by D. It is by convention that D is defined as in (3.4a), with a plus-sign in front, which then leads to another plus-sign in (3.4d) and minus-signs in (3.4b) and (3.4c). This notation could, of course, have been reversed. In general, it holds that population geneticists are normally much more interested in the *existence* of a D value different from 0 than in what sign – positive or negative – it takes. A non-zero D always indicates some functional, historic or random interaction between the investigated loci that carries interesting information about past evolutionary events.

D can be directly calculated if information is available about the allele frequencies and one of the haplotype frequencies. For example, from (3.4a), it follows that:

$$D = P_{11} - p(A_1)p(B_1) \tag{3.5}$$

However, D can also be expressed in terms of the haplotype frequencies alone. If (3.5) and (3.2a) and (3.2c) are combined, then we get:

$$\begin{aligned} D &= P_{11} - (P_{11} + P_{12})(P_{11} + P_{21}) \\ &= P_{11} - P_{11}^2 - P_{11}P_{12} - P_{11}P_{21} - P_{12}P_{21} = \\ &= P_{11}(1 - P_{11} - P_{12} - P_{21}) - P_{12}P_{21} = P_{11}P_{22} - P_{12}P_{21}. \end{aligned} \tag{3.6}$$

From this expression, it is seen that D can only take values between $-\frac{1}{4}$ and $+\frac{1}{4}$. In the former case, $P_{11} = 0, P_{12} = 1/2, P_{21} = 1/2$ and $P_{22} = 0$, while in the latter, $P_{11} = 1/2, P_{12} = 0, P_{21} = 0$ and $P_{22} = 1/2$. Observe that in these contrasting cases, the allele frequencies are the same – $p(A_1) = p(A_2) = p(B_1) = p(B_2) = 1/2$ – though their haplotype combinations are all different.

The change in D between generations

So far, we have introduced a succinct way to describe the relationship between allele frequencies and haplotype frequencies at two loci. Now, we turn to the question how any given linkage disequilibrium will change over time under random mating and recombination, in the absence of potentially disturbing factors such as selection and genetic drift.

We first note that if the diploid life-stage is formed through random fusion of gametes with frequencies (3.1a–d), then the diploid genotype frequencies will conform to the Hardy–Weinberg proportions for four alleles (see Background 2:1). Using this fact, we start by setting up a table giving the frequencies of all the possible diploid genotypes, including their gametic phase, and the gametes that they produce:

Diploid genotype	Frequency	Gametes produced			
		A_1B_1	A_1B_2	A_2B_1	A_2B_2
A_1B_1/A_1B_1	P_{11}^2	1	0	0	0
A_2B_2/A_2B_2	P_{22}^2	0	0	0	1
A_1B_2/A_1B_2	P_{12}^2	0	1	0	0
A_2B_1/A_2B_1	P_{21}^2	0	0	1	0
A_1B_1/A_2B_2	$2P_{11}P_{22}$	$(1-r)/2$	$r/2$	$r/2$	$(1-r)/2$
A_1B_1/A_1B_2	$2P_{11}P_{12}$	$\frac{1}{2}$	$\frac{1}{2}$	0	0
A_1B_1/A_2B_1	$2P_{11}P_{21}$	$\frac{1}{2}$	0	$\frac{1}{2}$	0
A_1B_2/A_2B_1	$2P_{12}P_{21}$	$r/2$	$(1-r)/2$	$(1-r)/2$	$r/2$
A_1B_2/A_2B_2	$2P_{12}P_{22}$	0	$\frac{1}{2}$	0	$\frac{1}{2}$
A_2B_1/A_2B_2	$2P_{21}P_{22}$	0	0	$\frac{1}{2}$	$\frac{1}{2}$

Let $P_{11}{}^*$ be the new value of P_{11} one generation later (this is, thus, yet another way to denote a change between generations). From the table, we find that:

$$\begin{aligned}
P_{11}{}^* &= P_{11}^2 + 2P_{11}P_{22}(1-r)/2 + 2P_{11}P_{12}/2 \\
&\quad + 2P_{11}P_{21}/2 + 2P_{12}P_{21}(r/2) \\
&= P_{11}^2 + P_{11}P_{22}(1-r) + P_{11}P_{12} + P_{11}P_{21} + P_{12}P_{21}r \\
&= P_{11}(P_{11} + P_{22}(1-r) + P_{12} + P_{21}) + P_{12}P_{21}r \\
&= P_{11}(1 - P_{22}r) + P_{12}P_{21}r \\
&= P_{11} - r(P_{11}P_{22} - P_{12}P_{21}) \\
&= P_{11} - rD.
\end{aligned} \tag{3.7}$$

From expression (3.4a), we know that we can always express $P_{11}{}^*$ as follows

$$P_{11}{}^* = p(A_1)^* p(B_1)^* + D^*,$$

where D^* is the linkage disequilibrium in the new generation.

[49]

Derivation 3

Recall, however, that in the absence of drift and selection, the allele frequencies $p(A_1)$ and $p(B_1)$ remain constant over time, which implies that $p(A_1)^* = p(A_1)$ and $p(B_1)^* = p(B_1)$, and that therefore

$$P_{11}^* = p(A_1)^* p(B_1)^* + D^* = p(A_1)p(B_1) + D^*. \qquad (3.8)$$

Combining (3.7) and (3.8) leads to

$$P_{11} - rD = p(A_1)p(B_1) + D^*,$$

which can be rewritten

$$D^* = P_{11} - p(A_1)p(B_1) - rD.$$

This gives us, using (3.5),

$$D^* = D - rD = (1 - r)D, \qquad (3.9a)$$

and the main result of this chapter has thereby been obtained.

Thus, one generation of random mating and recombination reduces the degree of linkage disequilibrium between the two alleles by a factor $(1 - r)$. From this, it follows that if the degree of linkage disequilibrium in an initial generation is D_0 and D_t is the linkage disequilibrium t generations later, then the following expression holds:

$$D_t = (1 - r)^t D_0. \qquad (3.9b)$$

The use of measure D to capture the relationship between alleles at different loci was initiated by Robbins (1918), and the key result (3.9) was first presented by him.

Our derivation shows that 0 can be regarded as an equilibrium point for the process given by expression (3.9a). Since $(1 - r)^t$ in expression (3.9b) always decreases with t, it is also clear that $D = 0$ is globally stable.

In a finite population, the exact value of $D = 0$ cannot be obtained and the linkage disequilibrium will be affected by random genetic effects. The expected distance from 0 is – of course – a decreasing function of the population size. The effect will normally be small, unless the loci are *very* closely linked (see Ewens, 1979).

Rate of change

A similar logic to that in Derivation 1 can be used to find how long it will take for recombination to reduce a particular D-value to half its current value. The number of generations to get there, $t_{1/2}$, is obviously given by

$$(1 - r)^{t_{1/2}} = \frac{1}{2},$$

which leads to

$$[e^{\ln(1-r)}]^{t_{1/2}} = e^{\ln(1/2)}$$

$$e^{t_{1/2} \cdot \ln(1-r)} = e^{[\ln(1)-\ln(2)]}$$

$$t_{1/2} \cdot \ln(1 - r) = \ln(1) - \ln(2)$$

$$t_{1/2} = \frac{0 - \ln(2)}{\ln(1 - r)}$$

$$t_{1/2} = \frac{\ln(2)}{-\ln(1 - r)}.$$

For free recombination between the loci, $r = 0.5$, the process will be very rapid (the half-life is exactly one generation), while with close linkage, say $r < 1\%$, the reduction in linkage disequilibrium will take many generations (for $r - 1\%$, the half-life is approximately 67 generations). Thus, if a significant linkage disequilibrium is found between two loci in a random-mating population, it is more or less certain that the loci are closely linked (*i.e.* in close physical proximity along a chromosome). Figure 3.1 illustrates how a linkage disequilibrium is broken down by recombination in a random-mating population where selection and drift can be ignored.

It is implicit in what we have written so far that the loci we consider are inherited via nuclear (autosomal) chromosomes. There may, of course, exist disequilibria between loci belonging to other kinds of genetic material, such as mitochondrial DNA or the sex chromosomes, but we do not discuss them here; just as we do not analyse how the breakdown of linkage disequilibrium is hampered by factors such as inbreeding, partial asexuality or population subdivision. Let us just state the obvious: that when only a reduced number of recombinants are formed per generation, it

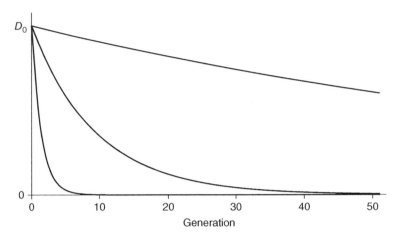

Figure 3.1 Decrease of an initial linkage disequilibrium, D_0, as a function of time and recombination in a large population and in the absence of other evolutionary factors (see expression (3.9b)). The dynamics is given for three different recombination values, $r = 0.50$, $r = 0.10$ and $r = 0.01$, starting from below. While, in the first case, hardly no linkage disequilibrium is detectable after, say, 10 generations, the D value is not much decreased over this timespan when the loci are tightly linked.

becomes more likely that other evolutionary factors than recombination, such as selection, migration or genetic drift, also affect the level of linkage disequilibrium between loci.

Summing up

Recombination acts to break up associations that may exist between alleles at different loci in a genome. The linkage disequilibrium parameter, D, is a convenient measure of such non-random associations, and we have now formally shown how recombination steadily decreases any linkage disequilibrium between two loci (in the absence of other evolutionary forces). Our derivation is, as in earlier chapters, based on the assumption that the size of the population is so large that all gamete frequencies are equal to their expected values. The main result of our derivation is well-known and is often given in textbooks, though normally without any proof. As long as random mating holds and there is no selection, the calculations are equally valid for organisms with haploid and for those with diploid dominant life stages.

Interpretations, extensions and comments
The importance of linkage disequilibria

It is important to realize that in all the processes discussed in this chapter, it is only the *haplotype frequencies* that change between generations due to recombination, and not the *allele frequencies*. Thus, the Law of Constancy of Allele Frequencies, which we presented in the Introduction, and which says that the frequency of an allele is expected to change only if it is affected by any direct evolutionary force, holds also when considered in a multi-locus setting.

That there may be associations between alleles at different loci has been with population genetics since its start, but it took a long time for linkage disequilibrium to become an important topic in the field. This was, of course, due to the lack of relevant data. For a long time, population geneticists were happy if they had information about the variation at one or at most a few loci, and the chance that any of these loci would be closely linked was normally very small.

This situation changed with the development of new molecular methods for identifying genomic variation. In a brief note, Solomon and Bodmer (1979) pointed out that with the then-new use of restriction enzymes so many polymorphic sites in the human genome could be identified that most genes would be in reasonably close linkage to at least one of them. Important genetic variation could, thus, be mapped on to chromosomes and studied via proxy – by some in-itself unrelated variation that marked the interesting chromosome region and made it identifiable in, for example, pedigrees. With the further development of full-scale sequencing technology, this has, of course, become a trivial possibility, at least in theory. All large-scale association studies, for example, build on the idea that the markers studied are in themselves of no particular interest; their importance derives instead from the DNA regions they identify. The fact that chromosome regions with effects on important traits can be identified and followed in populations, families or breeding lines, even when the exact genetic causative mechanisms are unknown, has revolutionized modern genetics. In all such analyses, the notion of linkage disequilibria between separate sites in the DNA is absolutely fundamental. In the first part of this chapter, we concentrated on the historically and pedagogically most useful measure of linkage disequilibrium, D. Other measures of associations between alleles at different loci are sometimes more practical and have become important in the literature – thus, for the use of D', see Question 4 later in this chapter, while for the squared correlation measure R^2, see the end of this discussion section.

It is relevant to realize that every new mutation – say, a base-pair substitution in a specific gene – automatically starts in positive association with the other genetic variants existing on the same chromosome. There is thus a positive linkage disequilibrium between the new allele and all other variable sites along the chromosome, such as single nucleotide polymorphisms (SNPs). Assume that the new mutation remains in the population for some significant time, be this due to chance or selection, or both. Its original linkage disequilibrium with the other variants along the chromosome will then gradually be lost due to recombination, more rapidly for distant sites and more slowly for close ones. Even after many generations, however, one would still expect there to be some positive linkage disequilibrium between this base-pair position and the markers close by – a situation found in all sequencing studies. To be perfectly correct, this argument builds on the key assumption that the new mutation is exactly that: new. If recurrent mutations occurs (*e.g.* if an original nucleotide T at a site repeatedly mutates to a G), then any disequilibria between this site and other sites nearby will decrease more rapidly than would happen from recombination alone.

Let us finally clarify one point of word usage. Assume that two polymorphic loci with several alleles each are studied. And let there exist an association, a significant linkage disequilibrium, between one allele at the first locus and one allele at the second. It is then standard practice to say that the two *loci* are in linkage disequilibrium, even if some alleles at the first locus are unassociated with some alleles at the second locus – a convention we follow. This phenomenon of alleles with and without linkage disequilibrium with one another can, however, never occur in the simplest possible situation of two loci with two alleles each. In such situations, it follows from the preceding discussion about degrees of freedom, and from (3.4), that if one of the alleles at one locus is in linkage association with one of the alleles at the other, then all three other allelic combinations will also be associated with one another.

Recombination and sex

The linkage disequilibria that may exist between different loci are not only of practical importance in biomedical research and in breeding programs based on marker assisted selection; they may also explain the ubiquitous presence of recombination and sex in eukaryotes.

It has always seemed reasonable that the long-term evolutionarily motivated function of regular sex and recombination is to produce hitherto unseen combinations of alleles at chromosomally linked loci. This explanation has, however, worried mathematical population geneticists from the beginning, since it turns out to be difficult to find a solid selective force favouring recombination between loci in well-characterized models – if anything, recombination tends in such models to become reduced (Feldman *et al.*, 1997)!

A convincing explanation for the process of recombination between genes had therefore to wait until Keightley and Otto (2006) presented data from large-scale computer simulations. These authors considered a structurally simple and robust situation, which nevertheless contained such richness that it could not be directly investigated analytically (which is a common way of saying 'with paper and pencil'). The assumptions they made were the following: A haploid organism has many genes along a chromosome at which moderately deleterious mutations constantly occur and are selected against. A brief diploid life-stage, formed by random fusion of the haploid cells, occurs every generation, during which recombination takes place between the loci along the chromosome, before the new haploid life-stage is formed. The population size is limited, though not necessarily small. Genetic variation exists in the population with respect to the amount of recombination between the loci under selection.

Keightley and Otto's (2006) computer runs convincingly showed that the rate of recombination between the loci will always tend to increase, assuming this set-up. This is because the selected removal of the deleterious mutations creates associations between beneficial and deleterious alleles at different sites along the chromosomes. Secondary selection then favours that such disequilibria are weakened, so that the nondeleterious variants may more easily become fixed. Importantly, this indirect selection favouring recombination does not depend on the studied population being small; instead, the selective strength favouring recombination increases for all realistic values on the population size.

Summarizing population situations: H and D

Population genetics has over the years introduced a number of useful concepts and measures with which the genetic variation in populations can be described and summarized. The linkage disequilibrium, D, between alleles at different loci is one of these, and we will later encounter F_{ST}, which describes the degree of genetic divergence between subpopulations. It is sometimes useful to know that these measures can be interpreted in a more statistical manner. To investigate this phenomenon, let us begin with the simplest case, one genetically variable locus.

When it comes to this variability, the number and frequency of the different alleles at the locus constitute the most basic information. It is, however, often useful to have a summary statistic for the amount of variability at the locus, and for this purpose the standard measure of *genetic diversity*, H, often called *heterozygosity*, is helpful. H is defined as the probability of getting different allelic types if two gene copies are drawn at random from the population. Thus, in a population with only two alleles, A_1 and A_2, having frequencies p and $1 - p$, the genetic diversity is $2p(1 - p)$. (This value also represents the expected frequency of heterozygotes A_1A_2 under random mating in diploids, since forming offspring under random mating can be seen as drawing two gene copies at random from the population's gene pool.)

Let us now assume a variable X which takes value 1 if a drawn gene copy from the population is of allelic type A_1 and value 0 otherwise. (If you need to refresh your memory on random variables, their expectations, variances and covariances, see Backgrounds 3:1 and 7:1). The expectation of this variable is obviously $p \cdot 1 + (1 - p) \cdot 0 = p$, the frequency of allele A_1. The variance of the variable is also easily found:

$$p[1 - p]^2 + (1 - p)[0 - p]^2 = p(1 - p)^2 + p^2(1 - p)$$
$$= p(1 - p)(1 - p + p) = p(1 - p).$$

We have now shown that this particular variance equals $1/2H$ (*i.e.* half the genetic diversity at the locus). Thus, the two measures are directly linearly related. In particular, they take the value 0 under the same (and trivial) circumstance of there being no variation at the locus.

We continue along this path with a similar study of D. Let us assume that there is a second locus with two alleles B_1 and B_2, having frequencies q and $1 - q$. When haplotypes are taken from the population, they have frequencies P_{11}, P_{12}, P_{21} and P_{22}, which makes the linkage disequilibrium, D, between the two loci $P_{11}P_{22} - P_{12}P_{21}$ (see (3.6)). In accordance with our method for the A locus, we assume that there is a random variable Y which takes value 1 if a drawn gene copy from the population is of allelic type B_1 and value 0 otherwise.

Now we want to know the covariance between the two variables X and Y. According to its definition, this value is:

$$Cov[X, Y] = P_{11}(1 - p)(1 - q) + P_{12}(1 - p)(0 - q) + P_{21}(0 - p)(1 - q)$$

$$+ P_{22}(0 - p)(0 - q)$$

$$= P_{11}(1 - p)(1 - q) - P_{12}q(1 - p) - P_{21}p(1 - q) + P_{22}pq$$

$$= P_{11} - P_{11}p - P_{11}q + P_{11}pq - P_{12}q + P_{12}pq - P_{21}p$$

$$+ P_{21}pq + P_{22}pq$$

$$= P_{11} - P_{11}p - P_{11}q - P_{12}q - P_{21}p$$

$$+ pq(P_{11} + P_{12} + P_{21} + P_{22})$$

$$= P_{11} - p(P_{11} + P_{21}) - q(P_{11} + P_{12}) + pq$$

$$= P_{11} - pq - qp + pq$$

$$= P_{11} - pq$$

$$= D.$$

In these calculations, we have used many of the assumptions and results from the derivation earlier in this chapter, ending with the key relationship (3.4a). We find that the linkage disequilibrium between loci may in a 2×2 allelic situation be identified with the covariance between the loci, defined in our specific way.

We can take this mode of reasoning one step further. The *correlation coefficient* between two random variables is defined as the covariance between the variables divided by the product of their respective standard deviations. In this case, the standard deviations of X and Y will not change over time (being $\sqrt{p(1-p)}$ and $\sqrt{q(1-q)}$ and thereby falling under the constancy of allele frequencies). Thus, a direct linear relationship exists between the linkage disequilibrium between the two loci and the correlation between the traits affected by the loci. When the linkage disequilibrium between the loci is 0, one can, thus, truly talk about them as being uncorrelated.

Sometimes the square of this correlation coefficient is used as a measure of the association between two biallelic loci; it is then generally denoted R^2. Thus, we have that:

$$R^2 = \frac{D^2}{p(1-p)q(1-q)}.$$

Ranging from 0 to 1, and being independent of the allele frequencies involved, this measure is sometimes more intuitively satisfying than D as a means of describing and summarizing how far from independence the variation at closely situated chromosomal sites is.

Background 3:1 Distributions – their expectations (means) and variances

Here, we give a brief summary of the most important results concerning distributions and how they can be characterized.

Let X be a discrete random variable that takes value x with probability $p(x)$, where x is one of the integers 0, 1, 2, 3, ..., and all p-values are such that $p(x) \geq 0$. The distribution is finite if it only takes values up to a fixed number and infinite if it can take all positive integers. To be a probability distribution, it must hold that

$$p(0) + p(1) + p(2) + p(3) + \ldots + p(x) + \ldots = \sum_x p(x) = 1,$$

where summation is made over all the x values the variable can take.

The *expectation* of the random variable X we denote $E[X]$, and it is given by:

$$E[X] = p(0) \cdot 0 + p(1) \cdot 1 + p(2) \cdot 2 + p(3) \cdot 3 + \ldots + p(x) \cdot x + \ldots$$
$$= \sum_x p(x) \cdot x. \tag{1}$$

This value is, of course, often called the (arithmetic) *mean*; a convenient shorthand for it is $\bar{x}$

If a is a constant, then $E[a] = a$, and if X and Y are two random variables, then $E[X + Y] = E[X] + E[Y]$.

A very useful result that we will use repeatedly is the following. If we are interested in the transformed variable which can be written $aX + b$, then its expectation is as follows:

$$E[aX + b] = aE[X] + b.$$

The result is self-evident, and its proof runs like this:

$$E[aX + b] = \sum_x p(x)(ax + b) = \sum_x ap(x)x + \sum_x bp(x)$$
$$= a\sum_x p(x)x + b\sum_x p(x) = aE[X] + b \cdot 1$$
$$= aE[X] + b.$$

A mathematically natural way to describe the spread of a variable is to find its *variance*, defined as follows:

$$V[X] = \sum_x p(x)(x - \bar{x})^2. \tag{2}$$

The variance is, thus, the expectation of the squared difference between the variable and its mean. A convenient result worth remembering is

$$V[X] = E[x^2] - E[x]^2, \tag{3}$$

or in nontechnical words: the variance is the mean of the squares of the distribution minus the square of the mean. This result can easily be proven as follows:

$$V[X] = \sum_x p(x)(x - x)^2 = \sum_x p(x)(x^2 - 2x\bar{x} + \bar{x}^2)$$

$$= \sum_x p(x)x^2 - 2\bar{x} \sum_x p(x)x + \bar{x}^2 \sum_x p(x)$$

$$= E[x^2] - 2\bar{x}\bar{x} + \bar{x}^2 \cdot 1 = E[x^2] - \bar{x}^2 = E[x^2] - E[x]^2.$$

Here, we use the convention that $E[x^2]$ denotes the expectation of the variable one gets from squaring the x-values.

If we want to know the variance of the transformed distribution $aX + b$, then we have that:

$$V[aX + b] = a^2 V[X]. \tag{4}$$

Again, the proof is very simple and relies on $E[aX + b] = aE[X] + b$:

$$V[aX + b] = \sum_x p(x)(ax + b - E[aX + b])^2 = \sum_x p(x)(ax + b - aE[x] - b)^2$$

$$= \sum_x p(x)a^2(x - E[x])^2 = a^2 \sum_x p(x)(x - E[x])^2 = a^2 V[X].$$

The results given here for the expectation and variance of a linearly transformed distribution will often be used in the later derivations. In Background 7:1, we write more about the properties of variances and discuss the role of *independence* for the variance of the sum of two variables.

The *standard deviation* of a random variable is the square root of its variance, often written just SD. If, say, the variable measures lengths in metres, then the mean and standard deviations are also in metres while the variance is in square metres.

Let us finally remember that for a continuously distributed random variable X with density probability distribution $f(x)$, the following two simple results hold:

$$\int f(x)dx = 1$$

and

$$E[X] = \int f(x)xdx. \qquad (5)$$

Here, the integration is to be made over the interval that is relevant for the distribution. These results will be of direct use to us later.

Questions

1. The four haplotypes A_1B_1, A_1B_2, A_2B_1 and A_2B_2 have frequencies of 0.62, 0.09, 0.16 and 0.13 in a particular population. There is random mating in the population, and the recombination fraction between the loci is 0.15. Calculate the linkage disequilibrium in the present generation, and then use the mating table presented at the beginning of the chapter to find the haplotype frequencies expected in the next generation. Based on these new haplotype frequencies, calculate the new value for the linkage disequilibrium in the population and the new allele frequencies.

2. When the haplotypes in the starting generation of Question 1 combine to form diploids, two types of double heterozygotes occur: A_1B_1/A_2B_2 and A_1B_2/A_2B_1. Find the frequencies for these different types. Do the same for the next generation, as calculated in Question 1. To what value will the difference between these two kinds of heterozygotes evolve over time, do you think? Can you prove your guess?

3. Consider two loci: A, with alleles A_1 and A_2, and B, with alleles B_1 and B_2, where $p(A_1) = 0.6$ and $p(B_1) = 0.7$. What is the numerically largest linkage disequilibrium (D_{max}) possible between these loci, given the allele frequencies?

4. Another common measure of association between alleles at different loci is D', which is defined as D/D_{max}, where D_{max} is defined as in Question 3. The maximum value of D' is, obviously, 1. If you know D' in one generation, what value do you expect it will take in the next (in the absence of factors other than recombination)?

5. Two populations are completely genetically distinct at two loci. A number of generations ago, they gave rise to a hybrid population when 60% of the ancestors moved in from the first population and 40% from the second. This new population has since been isolated, has practised random mating and has been without any noticeable random drift or selection. The recombination frequency between the two loci is 8%. Today, the linkage disequilibrium between the loci in this population is 0.09. How many generations ago was the population formed?

Derivation 4
Time to coalescence

Let us now become more realistic. And also more relevant with respect to the kind of sequence information that nowadays is commonly available.

In the first three derivations, we have assumed that the size of the considered population is very large. This is a perfectly acceptable approximation for many purposes. In physics, for example, thermodynamics is normally taught under the assumption that there are very many – in effect, infinitely many – independently moving molecules in the investigated systems. The problem with this type of approach in our biological context is, however, that most populations are not *very* large, and – as we shall see – many important properties depend on this fact. In the next three derivations, we will therefore treat phenomena that follow from the finite nature of real populations. How the amount of variation in a population is built up by an interplay between mutation and genetic drift is studied in Derivation 5, while Derivation 6 investigates the way selection and drift jointly affect the probability of fixation of new mutations.

We start, however, by concentrating on the most fundamental property that follows from populations being finite in number, namely that all individuals in them are related to one another. Or, expressed more precisely in genetic terms: all gene copies at a locus in a current population must be derived from a single gene copy, somewhere back in history.

It is immediately clear – even if we limit ourselves to the study of just two gene copies – that it is almost never possible to describe the exact inheritance history of a set of gene copies back to their common origin. Such an achievement would require detailed genetic information about every

Understanding Population Genetics, First Edition. Torbjörn Säll and Bengt O. Bengtsson.
© 2017 John Wiley & Sons Ltd. Published 2017 by John Wiley & Sons Ltd.

member in earlier generations of the population, of a kind that hardly ever is available. To gain insight into the origin of two or more gene copies, suitable simplifying assumptions must therefore be made about the history of the population; the tools of theoretical population genetics must also be made to work backwards within this assumed framework. The key property that we look for is the mean time in generations back to the joint origin of the studied gene copies. In the derivations that follow, we describe how this number can be found in an idealized type of population, called a *Wright–Fisher population*. After first having obtained a simple and satisfying result for two gene copies taken from a diploid population, the analysis is extended to cover the historical structure of three or more gene copies.

The size of the population from which the gene copies are taken and the breeding behaviour that is practiced by the individuals within it are at the centre of interest for the present chapter. The difference between haploid and diploid genetic material is also given special attention, and we finish with a discussion of how real populations can be made to conform – via the notion of an 'effective population size' – to the abstract population model used in the derivations.

In the next chapter, we shift our attention to the gene copies themselves, and we will there consider them as sequences with many mutable sites. Some of what here may appear dry and abstract will then find its use and meaning, when we describe how the sequence variation in a population is determined by the combination of mutation and genetic drift.

Analysis

Assumptions

Assume a population of diploids with the following simple breeding system: The size of the adult population is N, which is a fixed number, and this has always been the size of the population back in time. All individuals produce an infinite number of haploid gametes (using Mendelian segregation), and these unite at random to form new diploids, of which N develop into adults. This means that a single gene copy in one generation may give rise to zero, one, two … up to all $2N$ gene copies in the next adult generation. Of great importance is the fact that some of the gene copies in a generation will normally not be represented in the next generation, which – in the reverse – implies that the gene copies in a present-day generation will only represent a subset of the copies in the preceding generation.

We also assume that there is no selection based on genotypes during either the haploid or the diploid phase (thus, that all genetic variation is neutral). A population with these properties is commonly called a 'Wright–Fisher population'; see more about this later.

Since the gametes unite at random, there is a possibility of self-fertilization – that two gametes from the same individual will fuse to form a new diploid individual in the next generation. This is a sometimes unnatural effect that comes with the use of such a very simple model of reproduction. How our results can be made to fit also more realistic breeding systems must therefore be discussed separately, and we do so at the end of the chapter.

Finding the formula for coalescence of two gene copies

Let us start by considering two gene copies in the present generation of a diploid Wright–Fisher population of size N. The probability that they both originate from one and the same gene copy in the earlier generation (that they 'coalesce one generation ago', as it is called) is, obviously, $1/(2N)$ – since this is the probability that the second gene copy derives from the same template in the earlier generation as the first gene copy. The probability that they derive from different gene copies in the earlier generation is therefore $1 - 1/(2N)$.

Similarly, the probability that the two studied gene copies derive from different gene copies one generation ago but are copies of a single copy *two* generations ago (that they coalesce two generations ago) is $[1 - 1/(2N)][1/(2N)]$. This formula just says that what happens during one transmission between generations is statistically independent of what happens during any other transmission step.

In general, the probability that our two copies derive from a single gene copy exactly x generations ago and have been separate in their historical paths since then must therefore be $[1 - 1/(2N)]^{x-1}[1/(2N)]$.

[65]

Derivation 4

The mean and the variance for the time to coalescence

We have now obtained a discrete distribution describing the probability that the coalescence of two gene copies occurred x generations ago. If we denote $1/(2N)$ with p, coalescence x generations ago has a probability that may be written $p(1-p)^{x-1}$. These probabilities add up to 1 when summed over all possible x values, just as they should. This can be seen as follows, using the formula for the geometric sum S_0 from Background 4:1 (and the fact that $0 < p < 1$):

$$\sum_{x=1}^{\infty} p(1-p)^{x-1} = p\sum_{x=1}^{\infty}(1-p)^{x-1} = p\left[\frac{1}{1-(1-p)}\right] = p \cdot \frac{1}{p} = 1.$$

This probability distribution is called the *geometric distribution* (see Background 4:2). It is applicable to many situations where one waits for the first occurrence of something to happen; it has already been used indirectly in Derivation 1, when we discussed for how long a gene copy will stay in its current allelic state before mutating.

To find the mean time to coalescence for our analysed gene copies, we use our knowledge that the probability that they coalesced x generation ago is $[1 - 1/(2N)]^{x-1}[1/(2N)]$. In general, the expectation of a discrete distribution is

$$E[X] = p(0) \cdot 0 + p(1) \cdot 1 + p(2) \cdot 2 + p(3) \cdot 3 + \ \ldots\ + p(x) \cdot x + \ldots$$

$$= \sum_{x} p(x) \cdot x,$$

which in the present case translates into

$$0 \cdot 0 + [1/(2N)] \cdot 1 + [1/(2N)][1 - 1/(2N)] \cdot 2$$

$$+ [1/(2N)][1 - 1/(2N)]^2 \cdot 3 + \ldots$$

$$= \sum_{x=1}^{\infty} x[1 - 1/(2N)]^{x-1}[1/(2N)].$$

Again, we use the notation $p = 1/(2N)$, and can succinctly write the mean time to coalescence as $E[x] = \sum_{x=1}^{\infty} xp(1-p)^{x-1}$.

Sum S_1 from Background 4:1 now becomes convenient, and we find that

$$E[x] = \sum_{x=1}^{\infty} xp(1-p)^{x-1} = p\sum_{x=1}^{\infty} x(1-p)^{x-1} = p\frac{1}{[1-(1-p)]^2}$$

$$= \frac{p}{p^2} = \frac{1}{p},$$

or, quite simply, by resubstitution that

$$E[x] = \frac{1}{1/(2N)} = 2N. \tag{4.1}$$

We have hereby reached the fundamental – and beautiful! – result that the mean time in generations to coalescence for two randomly drawn gene copies in a diploid Wright–Fisher population is twice the size of the population. Which is the same as saying that it is equal to the number of gene copies in the population.

What about the variance of this time distribution? This result can again be reached quite simply by first noting – with the help of sum S_2 from Background 4:1, and continuing to write $1/(2N)$ as p – that

$$E[x^2] = \sum_{x=1}^{\infty} x^2 p(1-p)^{x-1} = p\sum_{x=1}^{\infty} x^2(1-p)^{x-1} = p\frac{1+(1-p)}{[1-(1-p)]^3}$$

$$= \frac{2-p}{p^2}.$$

The convenient formula for the variance presented in Background 3:1 then becomes useful; together with (4.1), it gives

$$V[x] = E[x^2] - E[x]^2 = \frac{2-p}{p^2} - \frac{1}{p^2} = \frac{1-p}{p^2}.$$

Thus, we have found that

$$V[x] = \frac{1-p}{p^2} = \frac{1-(1/2N)}{(1/2N)^2} = 2N(2N-1) \approx 4N^2, \tag{4.2}$$

when N is reasonably large.

[67]

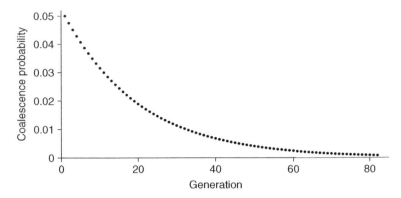

Figure 4.1 The probability distribution of time to coalescence, measured in generations, of two randomly drawn gene copies from a population of ten diploid individuals. The median of the distribution at 13.5 is much smaller than the mean at 20 (=2*N*; see expression (4.2)), due to the long tail and the asymmetric shape of the distribution.

It is well known that the geometric distribution has a wide spread, and this is illustrated here by the mean and the standard deviation (*i.e.* $SD = \sqrt{V} = \sqrt{4N^2} = 2N$) being approximately equal.

An example of a distribution of coalescence times is given in Figure 4.1. All the properties that we have just discussed can be seen there.

The coalescence history of more than two gene copies

Going from n *to* n − 1

The reader who just wants to know and understand what coalescence means and how this notion is handled in population genetics has probably learnt enough and can go directly to the section that discusses coalescence in haploid genetic material. The results produced in the next two sections are, however, of great importance for all theoretically based analyses of sequence data, and the road to achieving the results is not particularly difficult.

Instead of concentrating on just two gene copies, let us now study a sample of *n* gene copies taken from the present generation (we assume that the sample size is much smaller than the size of the population, *i.e.* that

$n \ll N$). We start, as before, by considering what may happen when one goes back one step in generations. Obviously, from 0 to $n - 1$ coalescence events may then occur. If 0 events occur, then the n gene copies derive from exactly n independent gene copies in the earlier generation. If $n - 1$ events occur, then all the n gene copies derive from a single copy in the earlier generation.

The probability for 0 coalescence events can easily be found by extending the logic used earlier. The first sampled gene copy derives from one of the copies in the earlier generation. The probability that the second sampled copy is derived from another gene copy than the first is $1 - 1/(2N) = (2N - 1)/(2N)$, just as before. The probability that the third sampled copy is derived from yet another gene copy than either of the first two is $1 - 2/(2N) = (2N - 2)/(2N)$. And so on. The probability that all n gene copies are derived from independent copies in the earlier generation is, thus,

$$P(0 \ coalesences) = \left[\frac{2N - 1}{2N}\right] \left[\frac{2N - 2}{2N}\right] \cdots \left[\frac{2N - (n - 1)}{2N}\right].$$

This can be rewritten as

$$P(0 \ coalesences)$$

$$= \left[1 - \frac{1}{2N}\right] \left[1 - \frac{2}{2N}\right] \left[1 - \frac{3}{2N}\right] \cdots \left[1 - \frac{n - 1}{2N}\right]$$

$$= 1 - \left[\frac{1}{2N} + \frac{2}{2N} + \frac{3}{2N} + \cdots + \frac{n - 1}{2N}\right] + \frac{C_1}{(2N)^2}$$

$$+ \frac{C_2}{(2N)^3} + \cdots + \frac{C_{n-2}}{(2N)^{n-1}}$$

$$= 1 - \frac{n(n - 1)}{4N} + \frac{1}{(2N)^2}\left(C_1 + \frac{C_2}{2N} + \cdots + \left[\frac{C_{n-2}}{(2N)^{n-3}}\right]\right),$$

where we use the standard formula for the sum of an arithmetic sequence (remind yourself of this if you don't remember it!) and we let C_1, C_2, ... C_{n-2} stand for the negative or positive constants that depend on n but not on N. As a consequence, we now know that the probability that at least one coalescence event occurs in the studied shift in generations must be:

$$p(1 \; \textit{or more coalescence events})$$

$$= 1 - p(0 \;\textit{coalescence events})$$

$$= \frac{n(n-1)}{4N} - \frac{1}{(2N)^2}\left(C_1 + \frac{C_2}{2N} + \cdots + \left[\frac{C_{n-2}}{(2N)^{n-3}}\right]\right).$$

A standard way to express a result like this is to say that the probability equals $\frac{n(n-1)}{4N}$ plus (or minus) terms of size $\frac{1}{N^2}$ (which means that these terms can be ignored if N is large).

What we next would like to know is $p(1 \;\textit{coalescence event})$, and to reach this value we will use an indirect approach. Let us first find the probability for exactly two coalescence events occurring during the studied generation step! There are many different ways for this to happen – let us focus on the case where the first three sampled gene copies derive from a single copy in the earlier generation, while all the other gene copies are independently derived. The probability for this event is obviously

$$\frac{1}{2N}\frac{1}{2N}\left[1 - \frac{1}{2N}\right]\left[1 - \frac{2}{2N}\right]\left[1 - \frac{3}{2N}\right]\cdots\left[1 - \frac{n-3}{2N}\right],$$

which we can rewrite as

$$\frac{1}{4N^2} \;\text{plus terms of size}\; \frac{1}{N^3}.$$

All the different ways by which exactly two coalescence events can occur have, in fact, the same probability of happening as the one just calculated (check this yourself!). The probability that two coalescence events occur in the transmission between generations can, thus, be written as $\frac{D}{N^2}$ plus terms of size $\frac{1}{N^3}$, where D is a constant that depends on n but not on N.

When it comes to three or more coalescence events during the studied generation transmission, it is easy to see that their respective probabilities become smaller with the number of assumed events. Thus, the probability for, say, five coalescence events will be an expression where the greatest term is of size $\frac{1}{N^5}$.

For n clearly smaller than N, and for substantial values on N, we can therefore approximate the probabilities for the different coalescence events as follows:

$p(n-1$ *coalescence events*$)$

$\approx \ldots \approx p(3$ *coalescence events*$) \approx p(2$ *coalescence events*$) \approx 0,$

$p(1$ *coalescence event*$)$

$= p(1$ *or more coalecence events*$) - p(2$ *or more coalescence events*$)$

$$= \frac{n(n-1)}{4N} - \frac{1}{(2N)^2}\left(C_1 + \frac{C_2}{2N} + \cdots + \left[\frac{C_{n-2}}{(2N)^{n-3}} \right] \right) - 0$$

$$\approx \frac{n(n-1)}{4N}, \text{ and}$$

$p(0$ *coalescence events*$) = 1 - \dfrac{n(n-1)}{4N}.$

Let us summarize: When a sample of n gene copies are studied from a population of N diploids, where N is reasonably large and n is small relative to N (two conditions that almost always are fulfilled), then we only need to consider the possibilities of there being zero or one coalescence event per generation. And the probability that the n gene copy histories decrease to $n-1$ histories due to such a coalescence event equals $\frac{n(n-1)}{4N}$.

From this simple formula, we see that the larger the sample, the more likely a coalescence event in the last generational step, just as expected. We see also that the probability for coalescence of two gene copies (*i.e.* $n = 2$) in this step equals $\frac{1}{2N}$, as it should according to our initial calculations.

Going from n *to 1*

In order to describe the whole historical process of coalescence, starting with n gene copies and going back until their lineages coalesce in a single copy, the concept of remaining linages is useful. If one starts with a sample of n lineages and goes backwards through generations, then, when the first coalescence event occurs, the number of remaining lineages decreases to $n-1$, and it is as if the coalescence analysis starts again at that moment,

[71]

but now with a sample of $n - 1$ gene copies. In this way, the analysis continues until the last two remaining lineages coalesce in a single unique copy. The whole inheritance history of the original n gene copies has then been followed from beginning to end.

We found earlier that if we have k remaining lineages, then the probability that they decrease to $k - 1$ when we go one generation step back is $\frac{k(k-1)}{4N}$, while the probability of no change is $1 - \frac{k(k-1)}{4N}$. We thus have here a new geometric distribution with parameter $p = \frac{k(k-1)}{4N}$ that describes the step-wise decrease in remaining coalescence lineages. With the same logic as used in the analysis of two gene copies, this distribution can be used to calculate the expected time during which there are k remaining lineages in the process. To simplify the description, let T_k denote this time. From the formula of the mean of a geometric distribution, we then know that

$$E(T_k) = \left[\frac{k(k-1)}{4N} \right]^{-1} = \frac{4N}{k(k-1)}. \tag{4.3}$$

Again, we see that this is compatible with our earlier result for $k = 2$, since the time to coalescence for two gene copies is $\frac{4N}{2(2-1)} = 2N$.

We can now proceed to calculate the expected time for the whole process when n sampled gene copies go through their inevitable coalescence events backwards in generations until one single gene copy remains. This is obviously the time it takes for the process to go from n gene copies over $n - 1$ remaining lineages, over $n - 2$ remaining lineages and so on, until the final coalescence event occurs when the last two remaining lineages come together into what we will call the most recent common ancestor of the studied sample. Thus, we have:

$$E[T_{mrca}] = E[T_n + T_{n-1} + \dots + T_2]$$

$$= E[T_n] + E[T_{n-1}] + \dots + E[T_2]$$

$$= \frac{4N}{n(n-1)} + \frac{4N}{(n-1)(n-2)} + \dots + \frac{4N}{2 \cdot 1}$$

$$= 4N \sum_{k=2}^{n} \frac{1}{k(k-1)} = 4N \sum_{k=2}^{n} \left[\frac{1}{k-1} - \frac{1}{k} \right]$$

$$= 4N \left[\left(\frac{1}{1} - \frac{1}{2} \right) + \left(\frac{1}{2} - \frac{1}{3} \right) + \dots + \left(\frac{1}{n-1} - \frac{1}{n} \right) \right]$$

$$= 4N \left(1 - \frac{1}{n} \right). \tag{4.4}$$

With this, we have found another beautiful formula, now for the expected time for the coalescence of n gene copies. We see that this time increases from $2N$ for two copies towards $4N$ with an increasing number of gene copies in the studied sample.

The coalescence tree and the length of its branches

Let us derive one more interesting measure for this coalescence process.

The coalescence history of n gene copies can be envisaged as a tree with n branches in the present generation, which become $n - 1$ when we go back in time, which become $n - 2$, and so on until the final two branches fuse into one. What is the total length of all these branches (measured in time given by shifts in generations)?

The total length of the branches when there are n of them is obviously nT_n, while it is $(n - 1)T_{n-1}$ when there are $n - 1$ remaining lineages, and so on. Thus, we have that the total length of the tree for n sampled genes, L_n, is the sum of the time spend with n remaining lineages plus the time with $n - 1$ remaining lineages, plus ..., which we can write as $L_n = nT_n + (n - 1)T_{n-1} + \dots + 2T_2$. Accordingly,

$$E[L_n] = \sum_{k=2}^{n} k \cdot \left[\frac{4N}{k(k-1)} \right] = 4N \sum_{k=2}^{n} \left[\frac{1}{k-1} \right] = 4N \sum_{k=1}^{n-1} \left[\frac{1}{k} \right]. \tag{4.5}$$

This number (for which no simple closed form exists) starts at $4N$ when two gene copies are analysed – the tree consists then of two branches with an average length of $2N$ each – and increases without a bound with the number of analysed gene copies (though, after a while, only slowly).

Why do we care about this? Well, when a gene copy is transmitted from one generation to the next, there is always a chance that it is hit by a mutation. Thus, when n gene copies evolve over time from their most recent common ancestor, the number of mutations that occurs in their coalescence tree is expected to be directly proportional to the length of the branches in the tree, that is, to L_n. In the next chapter, we will see how result (4.5) can be used to test whether the genetic variation found in a sample of sequences conforms to their having evolved under neutral genetic drift and random mutation alone, or if the data indicate the existence of, for example, some selective force.

Coalescence for haploid genetic material

A number of interesting results concerning the past history of gene copies in diploid populations have now been produced, condensed into formulas (4.1) to (4.5). Let us finish this section of derivations with a comment on how coalescence thinking functions when applied to haploid genetic material.

Actually, everything becomes even simpler. In a haploid Wright–Fisher population of size N, there are N adult individual organisms, carrying one gene copy each. They all produce infinitely many haploid propagules, from which the N adult individuals in the next generation derive. No fitness differences exist, either in fertility or in survival.

Going backwards, the probability that two randomly drawn gene copies from such a population coalesce one generation ago is now, obviously, $1/N$. Similarly, if we work through all the other preceding calculations, we find that wherever $2N$ was written for the diploid case, the correct value is now N. We can thus directly translate our earlier results as follows:

In a haploid Wright–Fisher population of size N, the mean time to coalescence for two randomly taken gene copies is N generations. This is also the approximate value for the standard deviation of their time to coalescence. The mean time back to the most recent common ancestor of a sample of n gene copies is $2N\left(1 - \frac{1}{n}\right)$. The expected length of time in the coalescence tree for this sample is $2N\sum\limits_{k=1}^{n-1}\left[\frac{1}{k}\right]$.

It is important to remember that a population of eukaryotic organisms will normally contain genetic material with different inheritance rules and therefore with different coalescence properties. Thus, in a human population of size N, we expect that the autosomes will fairly closely follow the logic outlined earlier for a diploid Wright–Fisher population (bar the possibility of self-fertilization). At the same time, the mitochondrial DNA, inherited only via females, and the Y chromosome, inherited only via males, follow the logic of a haploid Wright–Fisher population, in each case of size $N/2$ (given an equal sex ratio).

Summing up

All real populations are finite in size, which implies that a set of gene copies drawn from a current population can always be traced back to a common gene copy one, two, or many generations ago.

We derive the relevant formulae that link the mean time to coalescence to the size of the population (assumed to be constant, not small, and without any selection or recombination); see expressions (4.1) and (4.4). We also find the formula for the mean branch length of the coalescence lineages (4.5). Our analyses relate to a diploid population of size N, but they work equally well for haploid genetic material (though with a trivial parameter change). These results provide the theoretical background to all contemporary sequence-based analyses of DNA variation, as further discussed in the next chapter.

Interpretations, extensions and comments

New thinking within a classical framework

One can say that coalescence analysis consists of doing population genetics backwards in time in a highly simplified population model. The assumptions that the analysis is based on were implicitly used by Fisher (1922) in an early assessment of the effects of genetic drift, but it was Wright (1931) who first specified them in a well-defined model – this is the reason why we write about the Wright–Fisher population model.

There is nothing radically new to population genetics in the coalescence analysis, and a key result like (4.1) might well have been discussed by Wright or any of his contemporary colleagues. It was, however, after the publication of molecular data showing variable degrees of similarity between homologous sequences, that the British mathematician John Kingman (1982) first described the process whereby n gene copies coalesce to a single ancestral gene copy. His work inspired a vigorous development of coalescence-based techniques for the analysis of genetic data (summarized *e.g.* in Hein *et al.*, 2005).

Coalescence, inbreeding and identity by descent

Coalescence-like thinking had, actually, already been used for decades in population genetics, though only for a very limited and particular purpose – the study of inbreeding.

The French mathematician Gustave Malécot introduced the term '*identical by descent*' to designate gene copies that are related to one another via a finite number of replications (his early work is most easily accessed via various reprints and translations of Malécot, 1948). He based his analyses on pedigree information, and it is from him that we have the standard way of calculating the degree of inbreeding of an individual via its pedigree – this measure tells how much genetic identity there is between the analysed individual's mother and father. To be inbred means, in Malécot's terminology, that at least certain parts of one's diploid genome are *homozygous by descent*; that is, that very similar chromosome stretches (because they both derive from a copy in a recent ancestor) have been inherited from the mother and the father.

Today, when DNA can easily be sequenced and analysed, Malécot's terminology and way of analysis have become at the same time more relevant and more problematic. Even if we disregard the general problem that *all* gene copies in fact must go back to a joint ancestral copy and therefore – in a technical sense – are identical, we still have the problem that two analysed gene sequences may be *very* similar – indicating that they derive from a gene copy not very far back in time – but not absolutely *identical* in the sense that their base sequences are precisely the same. A change in meaning has therefore occurred, and today 'identical by descent' (IBD) and 'homozygous by descent' (HBD) are normally defined operationally as relating to sequences that are so similar that they must be derived from a close ancestral gene copy. For a study of how stretches of sequence similarity in humans can be used to estimate regions of such identity, see, for example, McQuillan *et al.* (2008). According to their analysis, an average European – who does not know about any consanguineous matings in his or her direct ancestry – is homozygous by descent for up to 4 Mb. The information that may be gained from the structure of such identity by descent blocks will become an important question for population genetics in our near future.

Wright–Fisher populations and effective population size

Wright (1931) was explicit about the fact that the breeding pattern and population model he used in his calculations were so simple and general that very few organisms, at least among the more complex ones, would reproduce accordingly. He therefore wrote about some organisms as having an 'effective population size', and meant by this that a particular population had such a size and breeding pattern that in population genetics terms it behaved as if it had size N_e in a Wright–Fisher population (except that he didn't, of course, call his generalized population model by this name).

The idea that some natural diploid populations of current size N with respect to genetic drift behave as Wright–Fisher populations of effective diploid size N_e has become very productive. Let us give three examples of what this effective population size is for some more realistic situations (the formulas we introduce are not difficult to derive, but we do not do so here).

A population with random-mating males and females (and where self-fertilization is therefore impossible) will with respect to random drift behave as a Wright–Fisher population with effective population size

$$N_e = \frac{4N_m N_f}{N_m + N_f},$$

where N_m is the number of reproductive males per generation and N_f is the number of reproductive females (both assumed to be constant over time). From this formula, it can be seen that, for example, the largest effective size for a population consisting of $N = N_m + N_f$ animals is obtained when the sex ratio is equal (*i.e.* when $N_m = N_f$). This is, thus, also the situation when the role of genetic drift is minimized for a fixed total number of individuals in the population.

[77]

A numerical example illustrates this point. In a population of 100 animals, the effective population size is 100 if the sex ratio is equal, since we than have $N_m = 50$ and $N_f = 50$, which makes $N_e = \frac{4 \cdot 50 \cdot 50}{50+50} = \frac{10000}{100} = 100$. If, instead, the population constantly consists of 20 males and 80 females, then N_e becomes $\frac{4 \cdot 20 \cdot 80}{20+80} = \frac{6400}{100} = 64$, with a corresponding increase in the effect of genetic drift.

If the number of diploids in an otherwise regular Wright–Fisher population always shifts between, say, a small number N_s and a large number N_l, then the effective population size is:

$$\frac{1}{N_e} = \frac{1}{2}\left[\frac{1}{N_s} + \frac{1}{N_l}\right].$$

The effective population size is, thus, given by the harmonic mean of the alternating sizes, which implies that it is much closer to the smaller than the larger number. This illustrates the more general insight that genetic drift caused by occasional reductions in population size is not easily compensated for by other generations being uncommonly large.

The effective population size, which in some sense is a measure of the influence genetic drift has on the evolution of the population, is – of course – also affected by the variance in offspring number to its members. If this variance is σ_{Off}^2, then the formula for the effective population size can be written:

$$N_e = \frac{4N - 2}{\sigma_{Off}^2 + 2}.$$

We see that the effective population size increases, and that genetic drift decreases, when the members in the population produce increasingly similar offspring numbers – just as expected. This result is well known among plant-breeders, who in situations where they want to keep a population as variable as possible often practise a breeding system called 'single-seed descent', implying that every plant is allowed to produce exactly one seed to be used for sowing the next generation.

Many other results regarding effective population numbers have been produced over the years (see *e.g.* Crow and Kimura, 1970 for more examples). The concept was originally studied with respect to the rate at which genetic variation would be lost from a population if this loss were not balanced by new mutations (see Derivation 5), but the idea works equally well when used in connection with coalescence calculations (Hein *et al.*, 2005).

Background 4:1 Some useful sums

Let S be the sum of the *geometric sequence*, $1 + a + a^2 + a^3 + \ldots + a^k$ (where we assume that $a \neq 1$). Since aS equals $a + a^2 + a^3 + \ldots + a^k + a^{k+1}$, it immediately follows that $S - aS = 1 - a^{k+1}$, and that therefore

$$S(1 - a) = 1 - a^{k+1}.$$

Thus, we have that

$$S = \frac{1 - a^{k+1}}{1 - a}. \tag{1}$$

When $k \to \infty$ and $0 < a < 1$, this sum goes towards $\frac{1}{1-a}$, which we will designate S_0. Thus,

$$S_0 = \frac{1}{1 - a}. \tag{2}$$

Let us now instead consider the sum $S_1 = 1 + 2a + 3a^2 + 4a^3 + \ldots = \sum_{k=1}^{\infty} ka^{k-1}$. Under the same condition as before, $0 < a < 1$, it follows that

$$S_1 - aS_1 = 1 + 2a + 3a^2 + 4a^3 + \ldots - a - 2a^2 - 3a^3 - \ldots$$

$$= 1 + a + a^2 + a^3 + \ldots = \frac{1}{1 - a},$$

where for the last step we have used the formula for the sum of a geometric series. Thus, we have now derived that

$$S_1 = \frac{1}{(1 - a)^2}. \tag{3}$$

Going one step further (still assuming $0 < a < 1$), we define S_2 as the infinite sum $S_2 = 1 + 4a + 9a^2 + \ldots + k^2 a^{k-1} + \ldots = \sum_{k=1}^{\infty} k^2 a^{k-1}$. Using the same procedure as before, we find that

$$S_2 - aS_2 = 1 + 4a + 9a^2 + \ldots + k^2 a^{k-1} + \ldots$$

$$- [a + 4a^2 + 9a^3 + \ldots + (k - 1)^2 a^{k-1} + k^2 a^k + \ldots]$$

$$= 1 + 3a + 5a^2 + \ldots + [k^2 - k^2 + 2k - 1]a^{k-1} + \ldots$$

$$= 1 + 3a + 5a^2 + \ldots + (2k - 1)a^{k-1} + \ldots$$

Thus,

$$S_2 - aS_2 = \sum_{k=1}^{\infty}(2k-1)a^{k-1} = 2\sum_{k=1}^{\infty}ka^{k-1} - \sum_{k=1}^{\infty}a^{k-1}$$

$$= 2S_1 - S_0 = \frac{2}{(1-a)^2} - \frac{1}{1-a} = \frac{2-1+a}{(1-a)^2} =$$

$$= \frac{1+a}{(1-a)^2},$$

and therefore,

$$S_2 = \frac{1+a}{(1-a)^3}. \tag{4}$$

Background 4:2 The expectation and variance of some common distributions

The proofs of the results given here can be found in any elementary text on probability theory.

Binomial distribution

The binomial distribution, often written as $Bin(n, p)$, is defined by two parameters, n and p, where n is an integer and p is a real number between 0 and 1. $Bin(n, p)$ gives the probability of outcome x, where x is an integer between 0 and n, as

$$f(x) = \frac{n!}{x! \ (n-x)!}p^x(1-p)^{n-x}.$$

A standard interpretation is the following: A trial is repeated n times, all independent of one another. Every time, a specific outcome – rather than its single alternative – may happen with probability p. Then, $f(x)$ gives the probability that this outcome occurs x times among the n events.

An example: If we throw a fair die three times, the probability that we get exactly two fours is

$$f(2) = \frac{3!}{2! \ 1!}(1/6)^2(5/6)^1 = 3\frac{5}{6^3} = \frac{5}{72}.$$

Here, we have set $p = 1/6$, $n = 3$, and we look for the probability that $x = 2$.

For the mean and the variance of a random variable X with probabilities described by $Bin(n, p)$, we have the following results:

$$E[X] = pn \tag{1}$$

and

$$V[X] = p(1 - p)n. \tag{2}$$

Geometric distribution

The geometric distribution is defined by a single parameter, p, which is a real number between 0 and 1. The distribution gives the probability of outcome x, where x is a natural number, as

$$f(x) = p(1 - p)^{x-1}.$$

A standard interpretation is the following: An event is independently repeated until a specific outcome occurs. Every time, the specific outcome may happen with probability p. Then, $f(x)$ gives the probability that this outcome occurs for the first time when the event has been repeated x times. An example: If we throw a fair die repeatedly, the probability that a six occurs at the third throw but not before is

$$f(3) = (1/6)(5/6)^2 = \frac{25}{216}.$$

Here, we have set $p = 1/6$ and we look for the probability that $x = 3$.

For the mean and the variance of a random variable X with probabilities given by a geometric distribution with parameter p, the well-known results are:

$$E[X] = \frac{1}{p} \tag{3}$$

and

$$V[X] = \frac{1 - p}{p^2}. \tag{4}$$

We have already used the result about the mean, when in the discussion of Derivation 1 we intuitively accepted that one must throw a die on average six times to get a specific number. Here, $p = 1/6$ and the mean of the distribution is, thus, 6.

The geometric distribution is highly skewed. Every consecutive probability is smaller than the one before, so $f(1)$ is always the largest probability. Nevertheless, for a reasonably small value on p, many repeats will normally have to be done before the specific outcome occurs.

Normal distribution

The normal distribution is given by the continuous function

$$f(x) = \frac{1}{\sigma\sqrt{2\pi}}e^{-(x-\mu)^2/2\sigma^2},$$

where μ is the mean of the distribution and σ^2 is its variance. The distribution extends from $-\infty$ to $+\infty$.

A common way to imagine the origin of a normally distributed trait in genetics is to see the trait as being influenced by 'infinitely many' factors, each with an 'infinitely small' positive or negative effect.

Questions

1. Consider a population of rabbits consisting of 2000 individuals on an isolated island. What is the expected coalescence time for two randomly drawn gene copies of (a) an autosomal locus, (b) a locus on the Y chromosome, and (c) a locus in the mitochondrial DNA? Assume discrete generations and no selection. (All mammals are diploid, and rabbits, like almost all other mammals, have an XX/XY sex determination system, leading to an equal sex ratio. Mitochondria are always inherited from the mother.)

2. If, for a long time, a breed of sheep is propagated by 10 rams and 400 ewes, what is the effective size of the population? What is the expected coalescence time for two randomly drawn gene copies of (a) an autosomal locus, (b) a locus on the Y chromosome, and (c) a locus in the mitochondrial DNA?

3. What is the effective number, N_e, of a diploid moth species with two generations per year, where the size of the summer generation is about 10^8 individuals and the size of the winter generation is only about 10^4 individuals?

4. A diploid population contains N individuals after having grown in size for many generations. Do you expect the time to coalescence of two randomly drawn gene copies from the current population to be greater or smaller than $2N$?

5. In this chapter, we derived how, for a sample of two gene copies, the probability of coalescence in one generation is $\frac{1}{2N}$ and the probability of no coalescence is $1 - \frac{1}{2N}$. Assume instead that we have a sample of three copies. There are then three possibilities for coalescence when going back to the previous generation: no coalescence event, one coalescence event, or two coalescence events (implying that all three gene copies originate from the same gene copy in the preceding generation). (a) What are the exact probabilities for these three events? (b) What are the probabilities of these three events if N^2 is so large that $\frac{1}{N^2}$ can be approximated by 0?

Derivation 5

The evolution of genetic diversity

The derivations in this chapter continue to describe the effect of biological populations being finite in size, even if potentially large in number. We start by finding the formula for how much genetic variation is expected in a diploid population where both mutation and genetic drift operate. The same population model is assumed as in the last chapter – the simple and well-defined Wright–Fisher model – though it is here used for going forward rather than backward in generations. We then extend our results by investigating how long it takes for a population to build up its standing genetic variation, and for how long the variation in a population will 'remember' a prior demographic disturbance to its current size.

In our derivations, we assume that every new mutation leads to a new allelic type. The biological background to this assumption is discussed at the end of the chapter, where we introduce the infinite sites model of genes and DNA stretches. Within the framework of this model, it is possible to analyse gene copies, not only with respect to them being different or not, but from the point of view of *how* different they are. When such information is available, powerful phylogenetic reconstructions can be made, and we sketch the basis for such work. We also discuss how sequence information can be used to test the processes that have led to the currently existing genetic variation in a population: Is the variation affected solely by random processes, or is there an indication that, for example, selection has also been involved?

The chapter finishes with a brief comment on the versatility of methods with which population genetic equations can be analysed.

Understanding Population Genetics, First Edition. Torbjörn Säll and Bengt O. Bengtsson.
© 2017 John Wiley & Sons Ltd. Published 2017 by John Wiley & Sons Ltd.

Analysis

Assumptions and definitions

Consider a population of diploid organisms with the same properties as the one studied in the last chapter: what is called a 'Wright–Fisher population'. Since generation 0, the population has in every generation been of size N. When the diploid individuals are reproductively mature, they produce infinitely many haploid gametes, which unite at random. Among the diploid fertilization products, N individuals remain when the next adult generation is ready to reproduce.

In this population, a locus is considered for which the mutation rate, μ, is so small that all terms of size μ^2 can be ignored. Mutations make gene copies non-identical, and all mutations are unique – the same allelic type can, thus, appear only once. All new alleles are also selectively neutral. The mutations are assumed to happen during the gametic phase.

The genetic composition in such a population will be determined by a balance between mutation and genetic drift. Mutation will constantly introduce new genetic variation into the population, while drift will cause variation to be lost by chance. The amount of genetic variation in the population will therefore be determined by its size and by the mutation rate; the configuration at the start of the population will also be important if it was founded not very long ago.

We study the genetic diversity in the assumed population by calculating the probability that two randomly drawn gene copies from the population are different, meaning that if we follow the copies back in time, then a mutation has occurred in at least one of the two lineages before they coalesce. We will also consider them different if they do not coalesce before the initial population in generation 0 is reached – thus, the starting generation is assumed to contain $2N$ different gene copies in its N diploid individuals. This probability of difference we denote H and call the population's *genetic diversity index* or *heterozygosity*. It has been introduced and discussed earlier in the book – see the section titled "Summarizing Population Situations: H and D" in Derivation 3.

From a practical point of view, it turns out that it is often simpler in the calculations to find a mathematical expression for the identity value

I, equal to $1 - H$, rather than for H directly. This is an approach that is often taken in population genetic analyses.

Recursion with drift and mutation

We start by considering a population that has existed for so long that all effects of its initial configuration have been obliterated by time. The balance between drift and mutation has then reached its equilibrium value, and it is this level of genetic diversity that we wish to determine.

In the first step, we ignore the probability of mutations happening during the last generational transition and use subscripts to denote generation number. It is then easy to see that

$$I_t = \frac{1}{2N} + \left(1 - \frac{1}{2N}\right) \cdot I_{t-1}.$$

Here, the first term to the right of the equal sign, $\frac{1}{2N}$, describes the probability that the two gene copies considered in generation t are identical because they both are copies of the same gene copy in generation $t - 1$. The second term describes the probability that the two copies derive from two separate copies in generation $t - 1$, but that they there — in their turn — were identical with probability I_{t-1}. We have, thus, been able to move the question of identity between the two gene copies one generation back in time.

Formally speaking, I_t is an *expectation* given I_{t-1}, which in its turn depends on earlier values, which depend on even earlier values, and so on ... However, to avoid clogging up the writing, we do not spell this out explicitly in the formulae.

The achieved relationship can also be expressed in terms of genetic diversity, $H (= 1 - I)$. We then have that

$$1 - H_t = \frac{1}{2N} + \left(1 - \frac{1}{2N}\right)(1 - H_{t-1}),$$

which can be simplified as follows:

$$1 - H_t = \frac{1}{2N} + 1 - \frac{1}{2N} - \left(1 - \frac{1}{2N}\right)H_{t-1}$$

$$H_t = \left(1 - \frac{1}{2N}\right)H_{t-1}.$$

[87]

With this, we have obtained the well-known result that genetic drift, in the absence of mutation, will steadily decrease the amount of genetic variation in a population. The expected rate of loss is given by $1 - 1/(2N)$. The result quantifies the fact, mentioned earlier, that not all gene copies in a generation will normally be represented in the next. (We remind the reader that t here marks forward-running time, while in the preceding chapter we reckoned time backwards in generations with the help of x.)

Now to the balancing force, mutation. The role mutation plays in models of this type is to introduce a possibility that two gene copies studied in generation t are different, despite the fact that they – according to the preceding logic – should be identical. This unexpected difference arises from one or both of the gene copies having mutated during the transition between generations. When we take this possibility into account, we get the following expression

$$I_t = (1 - \mu)^2 \left[\frac{1}{2N} + \left(1 - \frac{1}{2N} \right) I_{t-1} \right], \qquad (5.1)$$

which states that identity – as defined and used here – can only exist between the two studied gene copies if none of them mutated during the last step of transmission. Recall that we have assumed that all new mutations are unique; any new allele introduced by the mutation process in this step will therefore necessarily be different from all preceding alleles. The rest of our calculations in this chapter are based on this simple recursion formula, first explicitly described by Malécot (1948).

Equilibrium state

We expect there to be an equilibrium state which is reached when the relevant variable, I_t, has ceased to change between generations. To find this state, we use the same method of analysis as in Derivations 1 and 2 and set $I_t = I_{t-1} = I$, which leads to

$$I = (1 - \mu)^2 \left[\frac{1}{2N} + \left(1 - \frac{1}{2N} \right) I \right].$$

This expression can be simplified and transformed into a statement about the genetic diversity, H, instead of the genetic identity, I. Ignoring terms of size μ^2, we get

$$I = (1 - \mu)^2 \left[\frac{1}{2N} + \left(1 - \frac{1}{2N} \right) I \right]$$

$$I \approx (1 - 2\mu) \left[\frac{1}{2N} + \left(1 - \frac{1}{2N} \right) I \right]$$

$$2NI = 1 - 2\mu + (1 - 2\mu)(2N - 1)I$$

$$2NI = 1 - 2\mu + 2NI - I - 4N\mu I + 2\mu I$$

$$2NI - 2NI + I + 4N\mu I - 2\mu I = 1 - 2\mu$$

$$I(1 + 4N\mu - 2\mu) = 1 - 2\mu$$

$$I = \frac{1 - 2\mu}{1 + 4N\mu - 2\mu}$$

$$H = 1 - I = 1 - \frac{1 - 2\mu}{1 + 4N\mu - 2\mu} = \frac{1 + 4N\mu - 2\mu - 1 + 2\mu}{1 + 4N\mu - 2\mu}$$

$$- \frac{4N\mu}{1 + 4N\mu - 2\mu} \tag{5.2a}$$

It is common to assume that μ is so small that it can be ignored relative to 1, which leads to the standard result

$$H \approx \frac{4N\mu}{1 + 4N\mu}. \tag{5.2b}$$

The expected genetic diversity that follows from the balance between mutation and drift in a finite population has thereby been found – the main result of the present chapter. This was first presented in the current setting by Kimura and Crow (1964). It is seen that the expected value becomes small when $N\mu$ is small, and close to 1 when $N\mu$ is large. A graph illustrating expression (5.2b) is given in Figure 5.1.

It can be noted that if we call the expected number of new mutations per generation M, then $M = 2N\mu$ and expression (5.2b) can be written $H \approx \frac{2M}{1+2M}$. This shows that with much less than one mutation per generation (M small), the genetic diversity in the population will be small.

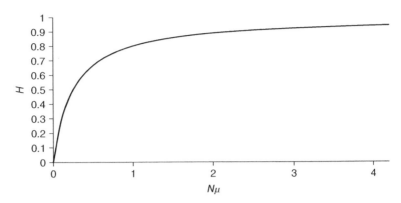

Figure 5.1 Expected genetic variability in a population, H, as a function of $N\mu$, the product of the population size and the mutation rate (see expression (5.2b)).

Again, it should be remembered that the calculated value is the *expected* diversity, and that, particularly in small populations, individual loci may well deviate far from this value. The variance associated with the expectation can, of course, be calculated, but we do not do so here.

Throughout this analysis, we have used N as the size of the diploid Wright–Fisher population studied. We might as well have written N_e to imply that the results will hold even for populations that do not exactly fit the assumptions of such populations but have the same drift properties as a Wright–Fisher population of size N_e. It has been shown that the notion of an effective population size, N_e, which we introduced and gave some examples of in the last chapter, works equally well here – and, furthermore, with exactly the same formulae.

Just as in the preceding chapter, all the derivations can easily be changed to cover haploid Wright–Fisher populations instead. The only difference is that the number of gene copies in the diploid case, $2N$, then corresponds to the number of individuals in the haploid population, N. Thus, the expected level of genetic diversity in a haploid population of size N is $\frac{2N\mu}{1+2N\mu}$.

The rate with which history is lost

For an analysis of the effect of the initial state (which we regard as generation 0) on the value of H in generation t, we return to expression (5.1), but proceed this time in a strict step-by-step manner. All derivations that follow, except for the very first, are exact and do not rely on any approximations. Self-evidently, the initial state does not have to be the point in time when the population was founded. More often, it can be thought of as the moment when the population reached the size it since has constantly kept. The mathematics is easy but tedious, and relies on working the recursion expression backwards. Much use is made of the sum, S, of a geometric sequence, as derived in Background 4:1.

Starting from (5.1), we have

$$I_t = (1-\mu)^2 \left[\frac{1}{2N} + \left(1 - \frac{1}{2N}\right) I_{t-1}\right]$$

$$\approx (1-2\mu) \left[\frac{1}{2N} + \left(1 - \frac{1}{2N}\right) I_{t-1}\right]$$

$$= (1-2\mu)\frac{1}{2N} + (1-2\mu) \left(1 - \frac{1}{2N}\right) I_{t-1}$$

$$= (1-2\mu)\frac{1}{2N} + (1-2\mu)^2 \left(1 - \frac{1}{2N}\right)\left[\frac{1}{2N} + \left(1 - \frac{1}{2N}\right) I_{t-2}\right]$$

$$= (1-2\mu)\frac{1}{2N} + (1-2\mu)^2 \left(1 - \frac{1}{2N}\right)\frac{1}{2N} + (1-2\mu)^2\left(1 - \frac{1}{2N}\right)^2 I_{t-2}$$

$$= (1-2\mu)\frac{1}{2N} \left[1 + (1-2\mu)\left(1 - \frac{1}{2N}\right)\right] + (1-2\mu)^2\left(1 - \frac{1}{2N}\right)^2 I_{t-2}$$

$$= (1-2\mu)\frac{1}{2N} \left[1 + (1-2\mu)\left(1 - \frac{1}{2N}\right)\right] + (1-2\mu)^3\left(1 - \frac{1}{2N}\right)^2 \frac{1}{2N}$$

$$+ (1-2\mu)^3\left(1 - \frac{1}{2N}\right)^3 I_{t-3}$$

$$= (1-2\mu)\frac{1}{2N} \left[1 + (1-2\mu)\left(1 - \frac{1}{2N}\right) + (1-2\mu)^2\left(1 - \frac{1}{2N}\right)^2\right]$$

$$+ (1-2\mu)^3\left(1 - \frac{1}{2N}\right)^3 I_{t-3}$$

$$= (1-2\mu)\frac{1}{2N} \sum_{k=0}^{t-1} (1-2\mu)^k\left(1 - \frac{1}{2N}\right)^k + (1-2\mu)^t\left(1 - \frac{1}{2N}\right)^t I_0$$

$$= \frac{(1-2\mu)}{2N} \frac{\left[1 - (1-2\mu)^t\left(1 - \frac{1}{2N}\right)^t\right]}{1 - (1-2\mu)\left(1 - \frac{1}{2N}\right)} + (1-2\mu)^t\left(1 - \frac{1}{2N}\right)^t I_0.$$

Derivation 5

Now, we change to studying genetic diversity, H, instead:

$$H_t = 1 - I_t = 1 - \frac{(1-2\mu)}{2N} \frac{\left[1 - (1-2\mu)^t\left(1 - \frac{1}{2N}\right)^t\right]}{1 - (1-2\mu)\left(1 - \frac{1}{2N}\right)}$$

$$- (1-2\mu)^t\left(1 - \frac{1}{2N}\right)^t(1 - H_0). \tag{5.3}$$

Before we proceed, let us first consider this expression for $t \to \infty$, and write $H_t \underset{t\to\infty}{=} H_{eq}$. Then, we have

$$H_{eq} = 1 - \frac{1-2\mu}{2N\left[1 - (1-2\mu)\left(1 - \frac{1}{2N}\right)\right]}, \tag{5.4}$$

which can be simplified as follows:

$$H_{eq} = 1 - \frac{1-2\mu}{2N - (1-2\mu)(2N-1)} = 1 - \frac{1-2\mu}{2N - 2N + 1 + 4N\mu - 2\mu}$$

$$= 1 - \frac{1-2\mu}{1 + 4N\mu - 2\mu} = \frac{1 + 4N\mu - 2\mu - 1 + 2\mu}{1 + 4N\mu - 2\mu}$$

$$= \frac{4N\mu}{1 + 4N\mu - 2\mu}.$$

This is exactly the result to expect, being equal to expression (5.2a), since this value describes the equilibrium balance between mutation and drift in a population that, after a long time, has gone to equilibrium. The formula for H_{eq} turns out to be useful when we describe the situation in a population that has not yet reached the equilibrium state. We return to our interrupted derivation (5.3) and start by using result (5.4):

$$H_t = 1 - \frac{(1-2\mu)}{2N} \frac{\left[1 - (1-2\mu)^t\left(1 - \frac{1}{2N}\right)^t\right]}{1 - (1-2\mu)\left(1 - \frac{1}{2N}\right)} - (1-2\mu)^t\left(1 - \frac{1}{2N}\right)^t(1 - H_0)$$

$$= H_{eq} + \frac{(1-2\mu)(1-2\mu)^t\left(1 - \frac{1}{2N}\right)^t}{2N\left[1 - (1-2\mu)\left(1 - \frac{1}{2N}\right)\right]} - (1-2\mu)^t\left(1 - \frac{1}{2N}\right)^t(1 - H_0)$$

[92]

$$= H_{eq} + (1 - 2\mu)^t \left(1 - \frac{1}{2N}\right)^t \left(\frac{1 - 2\mu}{2N \left[1 - (1 - 2\mu)\left(1 - \frac{1}{2N}\right)\right]} - (1 - H_0) \right).$$

Expression (5.4) can be used once more to give

$$H_t = H_{eq} + (1 - 2\mu)^t \left(1 - \frac{1}{2N}\right)^t [1 - H_{eq} - 1 + H_0]$$

$$= H_{eq} - \left[(1 - 2\mu)\left(1 - \frac{1}{2N}\right)\right]^t (H_{eq} - H_0).$$

By ignoring terms of size μ / N, we get

$$H_t \approx H_{eq} - \left[1 - 2\mu - \frac{1}{2N}\right]^t (H_{eq} - H_0), \tag{5.5a}$$

or, if we wish, with an extra, standard approximation (see Background 2:1):

$$H_t \approx H_{eq} - e^{-\left(2\mu + \frac{1}{2N}\right)t}(H_{eq} - H_0) = H_{eq} + (H_0 - H_{eq})e^{-\left(2\mu + \frac{1}{2N}\right)t}. \tag{5.5b}$$

Summing up

In finite populations, there is a steady loss of genetic variation due to genetic drift — which is another way of saying that not all gene copies in an earlier generation are transmitted to the next. When this effect is balanced by the introduction of new mutations, an equilibrium is expected. It turns out that the expected value of the resulting genetic diversity is given by a function determined by $N\mu$, the product between the population size and the mutation rate.

If the number of generations since the initial state, t, is larger than either the number of gene copies in the population ($2N$) or the inverse of the mutation rate ($1/\mu$), then the expected genetic diversity will be close to its equilibrium value, at least in principle, since the rightmost term in expression (5.5b) is then close to zero. Earlier than this, the effect of the initial condition will be noticeable, making the expected genetic diversity in the population higher or lower than the equilibrium value, depending on the initial state.

[93]

Interpretations, extensions and comments

Genetic variation, past demography and model assumptions

Our main result (5.2b) tells us that, for a constant mutation rate, the expected genetic diversity should be high in large populations and low in small populations. This theoretical result must, however, be interpreted with care when applied to more realistic situations than given by our model. Not just because the derived value is an expectation with a considerable variance (not calculated). But also because it takes a long time before the trace of an earlier population disturbance is lost, particularly if the population size is large. In many situations, when one deals with a large population, it is, of course, impossible to know whether it has had this large size for sufficiently many generations to be described by (5.2b). There are then reasons to look at equation (5.5) and consider whether demographic events that happened many generations ago might still exert some influence on the current level of genetic variation.

An interesting aspect of this was discussed by Nei *et al.* (1975) when they analysed what role a sudden bottleneck in size would have on the subsequent level of variation in a population. They pointed out that the dramatic loss of variation such a situation normally entails is due not to the primary event – the abrupt reduction in number – but to the subsequent limited number of individuals in the population when it starts to rebuild in size. A pregnant female, mated by many males, who with her offspring starts a new population will transmit a good proportion of the variation in the population she came from (a similar example, but referring to plants, is discussed in Question 3). However, if the newly founded population thereafter grows only slowly and therefore remains small for a number of generations, then it will come to carry only a very limited amount of variation.

All of this can also be understood with coalescence arguments. In an *expanding* population, the time to coalescence for a pair of gene copies is shorter than the current size of the population would have us believe, while in a *contracting* population the opposite holds. There will, in the first case, be less original variation to retain, while at the same time not many new mutations will occur during the relative short coalescence period.

[94]

Both of these aspects are reversed in the second case. A concrete example of this is given by the variability in humans and our closest living relatives. The level of genetic variation in our currently numerous species is comparatively low, while the levels of variation in the much rarer great primates are generally higher (Prado-Martinez *et al.*, 2013) – in crude terms, one can say that there is normally more genetic diversity in a troupe of Central African chimpanzees than in the whole of the human species. At least a part of the explanation for this paradox must follow from the fact that the chimpanzees retain a lot of variation from when their species was numerous, while anatomically modern humans represent the offspring of some few founders who gave rise to slowly growing populations that only recently have reached high numbers.

Genetic diversity, which usually is called *H*, is by convention instead designated π when applied to DNA sequence variation per base pair. An approximate but handy generalization is that π for humans equals about 0.001, which implies that if two homologous human DNA sequences are compared, then they differ at about 1 position in 1000. The exact value depends, of course, on the chromosomal position and on the population(s) from which the sequences come, but the quoted number gives the correct order of magnitude. This variation is primarily due to *single nucleotide polymorphisms* (SNPs), involving two base pair variants; how to best quantify the frequency of duplications and deletions in the genetic material (so-called *indels*), we leave to the technical literature.

Let us finally remember that result (5.2b) is derived based on specific assumptions. Most deeply problematic is the assumption that the variation at a locus can be analysed independently of what happens in the rest of the genome. This is only acceptable as a first approximation, since selection at other, and sometimes linked, loci will affect the way transmission occurs between generations, and thereby disturb the neat logic of the assumed Wright–Fisher model. Thus, it has been empirically found that the level of genetic variation in species of large size, where weak secondary effects also matter, is lower in the proximity of functional genes than elsewhere in the genome (Corbett-Detig *et al.*, 2015). Explanations of this phenomenon are discussed at the end of the next chapter.

The infinite sites and other mutation models

Mutations can be of different kinds. All are characterized by altering the DNA sequence, but the change may range from a single base-pair substitution at a particular DNA position to a large-scale reorganization of a whole chromosome. The different kinds of mutations are often associated with different mutation rates, which comes as an advantage to the analysing scientist, who can choose to work with the class of mutations that suits the investigated question best.

In empirically motivated population genetics, it is often relevant not to talk about mutations in an abstract way (as we do in our derivation), but to take their specific molecular nature into account — to incorporate a model of the particular mutation process into the analysis. In a study of microsatellite markers, for example, it is useful to assume that an allele with n repeats may mutate into alleles with $n - 1$ or $n + 1$ repeats, but not into other states. (Microsatellites differ in their number of short tandem repeats, typically in the range of $15-30$; for more on *microsatellite markers*, see any modern textbook in genetics.) Such a situation, where mutations occur as between steps in a ladder, is generally called the *stepwise mutation model*.

This model cannot be applied to the situation analysed in the present chapter's derivation, since we there assume that all mutations are unique and that they occur only once. (In the stepwise mutation model, this is not the case, since repeated mutations may flip a gene copy back and forth between two allelic states.) A more relevant mutation model, which also fits the structure of DNA very well, is then the *infinite sites model*, first outlined by Kimura (1969). The gene (or any well-defined part of a chromosome treated as a locus) is here considered to be a stretch of infinitely many sites (nucleotides). Mutations hit these sites at random, and every mutation creates a new and unique allele (since the sites are infinitely many, no position will be hit more than once by mutation).

This model corresponds closely to what genes actually look like at the DNA level. Since neutrality is often assumed in contexts where the model is used, the focus is restricted to the sites and mutations that are deemed to be neutral, or effectively so (and most genes contain a large number of such sites and mutations). For the model to function well in analyses of real gene samples, no site should be mutated more than once, and no intragenic recombination should occur within the gene – assumptions that are normally fulfilled when short divergence times are considered.

There is, actually, a very simple way to check whether a sample of sequences follows the logic of the infinite sites model. In such a situation, one should only find three haplotypes when two variable sites are considered (the haplotypes A_1B_1, A_1B_2 and A_2B_1, say, where A and B denote different sites along a studied sequences). This is because, if all four possible haplotypes are found, then either the same mutation must have happened more than once, or recombination must have occurred between the sites – how could the fourth haplotype, A_2B_2, otherwise have originated? Finding more than two nucleotide variants at a single DNA site would, of course, also indicate that a more sophisticated mutation model should be used.

Coalescence *and* mutations

We now have two sets of simplifying assumptions that may well be combined: the Wright–Fisher model of transmission over generations and the infinite sites model of gene mutations. These models are (reasonably) easy to study analytically and to implement in computer simulations, and together they describe what pattern of variation to expect in situations fulfilling the model assumptions.

Together, they help introduce an element that did not originally exist in population genetics, namely the idea that alleles are not only separable from one another, but that different alleles can be differently close or distant to one another. Two sequences that differ at many sites are, obviously, more distant from each other than two sequences with only few site differences. We will now discuss two situations where this knowledge is used to great effect, the first concerning phylogenetic reconstructions and the second relating to analyses of what processes have led to the particular pattern of genetic variation seen in a given sample.

Reconstructing phylogenies

Assume that we study three sequences, *a*, *b* and *c*, of some gene, taken from three different species, A, B and C. Assume also that we find that sequences *a* and *b* are more similar to each other than either of them are to sequence *c*. It is then reasonable to assume that *a* and *b* coalesce first when we go back in time before their remaining sequence coalesces with *c*, and that therefore the species tree should put A and B as being more related to each other than either is to C. This argument is widely used and perfectly correct, but it is worth discussing why and how a deduction like this sometimes leads to a factually erroneous result. (We only consider the logical structure of the argument here; all practical details are again to be found in the specialized literature.)

The first cause of error is due to the random nature of mutations. It could be that sequences *b* and *c* in fact coalesce first in backward-looking time, before their remaining lineage does so with *a*, but that this fact is hidden by unexpectedly many mutations having occurred in the lineage leading to *c* after its split from the lineage to *b* – all by pure chance. Since *c*, then, is more different from *a* and *b* than these are from each other, the reasonable – but erroneous – conclusion will be that A and B are more closely related to each other than either is to C. This type of problem can in many instances be identified by a closer analysis of the mutational pattern by which the sequences differ, since as long as the infinite sites model holds, it is always possible to work out the true pattern of divergence between sequences (given a trustworthy out-group sequence).

The second problem derives from the fact that a sequence coalescence tree (as discussed in the last chapter) does not have to fit exactly with the corresponding species branching tree. A discrepancy between them may arise if the genetic variation by which the sequences today differ existed *before* any of the evolutionary units started to separate. Our conclusion about the pattern of coalescence of the sequences may then indeed be correct, while also being misleading with respect to the species' phylogenetic branching pattern. In our case, this may, for example, happen if species A and B inherited more similar sequences than C, despite the fact that species A was the first to split from the joint ancestral population. This phenomenon, called incomplete lineage sorting, can normally be identified and rectified by using data from many separate and unlinked parts of the genome. Thus, nobody questions today the closer taxonomic relationship between humans and chimpanzees than between humans and gorillas, despite the fact that the gorillas are more closely related to humans than to chimps according to a fair proportion of possible sequence comparisons (Scally *et al.*, 2012).

These simple considerations explain why phylogenetic reconstruction from sequence data works best when the splits in the evolutionary tree were sufficiently well spaced in time to fit the mutational divergence of the genetic material involved (or, expressed in another way, for all relevant coalescence processes to occur within well-delimited taxa). It is also, in general, easier to establish the branching pattern of fairly small evolutionary units, since incomplete linage sorting occurs there less often (due to a lower level of variation in the original basal lineage).

Let us finally stress that the important conclusion we wish to present in this section is not that arguments based on the interactions between mutation and drift occasionally give incorrect predictions in phylogenetic reconstructions; instead, we wish to point out how fantastic it is that such analyses are at all possible, even with data coming from perhaps only a single gene. In classical population genetics, the analysis of a single gene would normally give information about two or at most a few allele frequencies and not very much could be done with this rudimentary knowledge. With sequence data, however, a reasonable-sized sample from a single DNA stretch provides a wealth of important evolutionary information by offering details about the similarities and differences *between* the sampled sequences.

Random processes produce discernible patterns

The sequences in a sample are also informative about the processes by which the observed variation has come about. Just because generative processes are based on the occurrence of random events does not mean that their end results will be without distinctive structures. Also random processes produce discernible patterns.

An interesting and powerful way to check whether an observed sample has been produced by some hypothetical process is – quite simply – to run a large number of computer simulations based on the assumptions using different parameter combinations and see if the observed pattern is compatible with the assumed model or not.

It is, however, often preferable – particular from a pedagogical point of view – if such model tests can be made in a simpler and more direct way. This is where the results from the later part of our derivations in the preceding chapter enter. Let us give an example of how they can be used to investigate whether a sample has been produced by the random processes assumed in the Wright–Fisher and infinite sites models.

We have a sample of sequences and we want to know whether their pattern of variation fits what could be expected from the model assumptions. To do this, we use two distinct ways to estimate the key quantity for our processes, $4N\mu$, and then compare the results.

On the one hand, we know that the mean time to coalescence for two gene copies in the sample is $2N$. This implies that the expected number of mutations by which the two copy sequences will differ is $2N \cdot 2 \cdot \mu = 4N\mu$, since mutations may occur in either of the two lineages leading to them. To get a better estimate of this number, we can calculate the mean of all possible pairwise differences between the sequences in our sample.

On the other hand, we know from result (4.5) in the preceding chapter that the mean total time of all the lineages back to their origin is $E[L_n] = 4N \sum_{k=1}^{n-1} \left[\frac{1}{k}\right]$ generations. The number of all mutations in our sample will therefore be $4Na \cdot \mu = a \cdot 4N\mu$, where $a = \sum_{k=1}^{n-1} \left[\frac{1}{k}\right]$; this number is directly observable from our sequences, since this is the number of sites at which variation occurs.

Thus, the number of segregating sites divided by a should equal the number of pairwise differences in our sample. Since we are dealing with stochastic processes, exact identity is, of course, not expected. Should, however, these values differ considerably from one another, then this is an indication that our set of sequences has not been produced by the posited assumptions.

Tajima (1989) developed a statistical test of the null hypothesis that these two estimates of $4N\mu$ are equal, which has become widely used. Its interpretation is, however, not straightforward. Natural selection affecting the variable sites, or sites closely linked, can, of course, produce almost any kind of nonrandom pattern and thereby affect the test statistic. From our earlier analyses, we understand that the population history may also play a role. To take but one example: If the size of the population was small until recently, when a dramatic increase took place, then there will be a smaller number of variants in the coalescent tree than would be expected under a large, constant population size, and few of the sequences will share the same variants. The coalescence tree will therefore look more like a bush – a so-called 'star phylogeny' – and the two measures of $4N\mu$ will be differently affected, giving rise – if the population expansion is sufficiently dramatic – to a negative and numerically significant test value.

Thus, one can say that whenever a significant value for a Tajima test is reached, no directly interpretable result has been produced, but an imperative for deeper analysis has been given. (All treatises on genetically based bioinformatics contain further discussions of these questions.)

Two ways of analysing the same equation

We end by making a technical comment about our main derivation.

Note that expression (5.5b) has a structure very similar to that of expression (1.5b), obtained in the analysis of the mutation–mutation balance, despite these expressions being reached in completely different manners. The structural relationship between the two cases becomes more evident with the following rewriting. Using expression (5.1) plus standard approximations, we get:

$$\Delta H = H_{t+1} - H_t = 1 - I_{t+1} - 1 + I_t = -I_{t+1} + I_t$$

$$= -(1-\mu)^2 \left[\frac{1}{2n} + \left(1 - \frac{1}{2N}\right)\right] I_t] + I_t$$

$$\approx -\frac{1}{2N} - I_t + 2\mu I_t + \frac{1}{2N} I_t + I = -\frac{1}{2N} + \left(2\mu + \frac{1}{2N}\right) I_t$$

$$= -\frac{1}{2N} + \left(2\mu + \frac{1}{2N}\right)(1 - H_t)$$

$$= 2\mu - \left(2\mu + \frac{1}{2N}\right) H_t = 2\mu - \left(\frac{1}{2N} + 2\mu\right) H_t.$$

Derivation 5

As seen, this expression is highly similar to expression (1.4),

$$\frac{dp_t}{dt} = \mu_{21} - (\mu_{12} + \mu_{21})p_t,$$

from which the derivation to reach (1.5b) was started. Structurally, they differ only in that the first is a difference equation in H, while the second is a differential equation in p. One can say that expression (5.5b) is reached after a rigorous derivation followed by a final approximation, while in expression (1.5b) an early approximation is instead used to turn a difference equation into a differential equation, which is then analysed without any further simplifications. Again, this illustrates that a good model in population genetics can be analysed in many ways and with different mathematical tools.

Questions

1. The mutation rate for a gene in a diploid population is $0.14 \cdot 10^{-6}$. The size of the population has been 10^7 for a very long time. An environmental collapse occurs, which, over one generation, reduces (stably) the population size to 10 000. What is the expected genetic diversity for the gene (a) before the collapse, (b) immediately after the collapse, and (c) 100, 1000 and 1 000 000 generations after the collapse?

2. A small founder population becomes established on a new continent, where it spreads and soon reaches a constant size of 10^6 individuals. When this size is first reached, the level of variation for the same gene as in Question 1 is still as low as 0.10%. What is this value expected to be after 100, 1000 and 1 000 000 generations?

3. A single head, containing many seeds, is taken from a very large population of an outbreeding, self-incompatible grass. How much of the population's genetic variability is retained by the seeds in the head?

4. Consider a locus, A, in a random-mating diploid organism. There are two alleles, A_1 and A_2, with frequencies p and q. Sample one individual from this population and regard it as a subpopulation (even if it only has one member). Show that the expected genetic diversity of this subpopulation, H_S, equals $H_T/2$, where H_T is the diversity in the original population.

5. Assume that an *allopolyploid* plant species has its origin in a single individual that lived many generations ago. Further assume that there is no variation among *chloroplast* DNA molecules within plants in the species, that the number of individual plants in the species grew very quickly up to a stable population number of 50 000, and that the mutation rate per base pair and generation for such DNA is $3 \cdot 10^{-9}$. What equilibrium genetic diversity is then expected for a chloroplast DNA sequence that is 10 000 base pairs long? How many generation will it take before the genetic diversity in the species/population reaches a level of 5%? (This question is inspired by the conditions in *Arabidopsis suecica*, as experienced by one of the book's authors – see Säll *et al.*, 2003.)

Derivation 6

Fixation of mutations with and without selection

We will now study how genetic drift affects the fate of mutations in finite-sized populations, concentrating in particular on the interaction between drift and selection. Our main derivation relies on a number of interesting mathematical techniques and leads to a second-degree differential equation that is solved with standard methods. We introduce some new mathematics under way, but also reuse a number of results presented earlier in the book. The going may seem a bit hard, but it is interesting to follow how a complex biological question can be mathematically tackled and ultimately lead to a simple and meaningful result. One important thing we learn is that a new beneficial mutation may easily be lost from a population during its very first generations due to simple chance effects.

This probability of loss is, of course, even greater for mutations without any direct effect on fitness: the so-called 'neutral mutations'. But not all of them will be lost. There is always a non-zero probability that a new neutral mutation will go to fixation by chance alone and without any systematic evolutionary force being involved. We derive another nice result about the rate of turnover of alleles at a locus where new neutral mutations constantly appear. And at the end, we see how fixation, variation and coalescence in combination may produce phenomena such as background selection and the hitch-hiking effect.

We are then back to questions about how the information in a set of DNA sequences can best be interpreted, as discussed in the previous

Understanding Population Genetics, First Edition. Torbjörn Säll and Bengt O. Bengtsson.
© 2017 John Wiley & Sons Ltd. Published 2017 by John Wiley & Sons Ltd.

chapter. The results in the present chapter are, however, not limited in importance to this particular, practical purpose. When we map the population genetic effects of drift and selection – ranging from positive over absent to negative – we let the material world of ecological interactions confront the equally material world of DNA's base pair sequence. And thereby, we come as close as possible to a general understanding of how the extrinsic world of an organism and its intrinsic molecular construction are coordinated in evolution.

Analysis

Assumptions and notations

Consider a population of constant size N of a diploid, random-mating organism. Further, consider a locus A, with alleles A_1 and A_2, which is subject to selection so that the relative fitnesses of genotypes A_1A_1, A_1A_2 and A_2A_2 are 1, $1 - s/2$ and $1 - s$, respectively. Thus, the allelic difference in question has a completely *additive effect* on fitness. This kind of fitness scheme has already been discussed in Background 2:3. It will further be assumed that s is a small and strictly positive number, plus that N is large, if not necessarily very large. With this, the population becomes similar to the diploid Wright–Fisher populations discussed in the two preceding chapters, but now with weak selection incorporated.

Selection acts in favour of allele A_1. Yet, since the population size is limited, genetic drift may come to fix A_2 instead. As a consequence, there has to exist a probability for the fixation of A_1, which we will denote $f(x)$, and which will depend on s, on N and – in particular – on x, the present frequency of allele A_1. This probability is certainly greater than zero but it is not necessarily equal to one. It will obviously increase with x, the frequency with which the positive allele A_1 already exists in the population. It is by finding this equation and seeing how it depends on the key parameters s and N that we get an understanding of how selection and genetic drift interact during evolution. In this derivation section, the original difference between A_1 and A_2 is the only genetic variation considered; the effects introduced by recurrent mutations are discussed later in the chapter.

Strategy

The purpose of our analysis is to find the function $f(x)$ and to analyse it for its implications. Our approach goes via first showing that f must satisfy the following differential equation:

$$2Ns\frac{df(x)}{dx} + \frac{d^2f(x)}{dx^2} = 0. \tag{6.1}$$

This equation is of a type commonly found in physics, particularly in connection with various diffusion phenomena, for example the spread of heat via molecular movement. We treat (6.1) as a mathematical result coming out of probability considerations, and we need to introduce some new mathematical techniques to be able to derive it from first principles. For alternative approaches to the same derivation, see Ewens (1979) and Gillespie (2004).

When we have convinced ourselves that the function describing the probability of fixation must satisfy (6.1), then we can rely on methods that we know from earlier and solve the differential equation. To find the specific function that we are interested in among all the solutions, we will in addition use the two obvious boundary conditions that $f(0) = 0$ and $f(1) = 1$, which say that an allele that doesn't exist in a population cannot go to fixation and that an allele with frequency 1 has already become fixed.

Deriving the differential equation

If the frequency of allele A_1 is x in a particular generation, then it will – most probably – not be *exactly* the same in the next generation, since we assume that it will be affected by positive selection as well as by stochastic chance effects. Let us write its frequency in the next generation as $x + \delta$, where δ is a random variable. In principle, the δ values will range from $-x$ over 0 to $1 - x$, but when the population size, N, is large and the selective advantage of the A_1 allele, s, is weak, the δ values will normally be numerically small. Let us also write the probability that the frequency is at this particular value as $p(x + \delta)$, where $p(x)$ – like $f(x)$ – is a continuous function defined for x ranging between 0 and 1. Since we are dealing with probabilities, the following must obviously hold

$$\sum_{\delta} p(x + \delta) = 1,$$

where the summation is made over all possible δ values (including the negative ones). Later, we will return and study these δ values more closely.

From the information currently available to us, we can now express the probability of fixation of the A_1 allele using what we know about the two considered generations. We have that:

$$f(x) = \sum_{\delta} p(x + \delta) \cdot f(x + \delta). \qquad (6.2)$$

This identity is the key to all our endeavours in arriving at expression (6.1). To the left is given the probability of fixation knowing the allele frequency in the present generation; to the right is given the probability of where the frequency will be in the next generation and the corresponding probabilities of fixation. We do not know 'for real' what the allele frequency will be in the next generation – if we did, then the probability of fixation would have changed. Instead, (6.2) is just a way to rewrite our knowledge about the present generation into the next. This may not seem much of an advance, but it is in the details of this rewriting – as will be seen – that the assumptions about drift and selection enter our calculations in an interesting way.

By going back to the definition of the expectation of a distribution (see Background 3:1), we can express (6.2) as

$$f(x) = E[f(x + \delta)], \qquad (6.3)$$

where it is implied that the expectation is to be taken over the distribution of all possible δ values.

To proceed further, we concentrate first on simplifying the $f(x + \delta)$ expression, and then on characterizing the distribution of δ values. In all these calculations, we regard the x value as given (constant); later, it will be treated as a variable, but not yet.

Simplifying f(x + δ)

We take for granted that the function $f(x)$ is continuous for all values of x in the interval 0 to 1, these values included; we also assume that the function 'behaves nicely' – which here technically means that it is differentiable infinitely many times at all points.

It is then possible to employ the very useful mathematical procedure that goes under the name of 'Taylor expansion'. In the present context, this implies that we can express the value of function f at the point $x + δ$ as a function of its behaviour at the point x – see Background 6:1. When we do this and ignore all terms containing cubes and higher powers of $δ$ (since $δ$ is small), we get:

$$f(x + δ) \approx f(x) + \frac{df(x)}{dx}δ + \frac{1}{2}\frac{d^2f(x)}{dx^2}δ^2.$$

This result, we use as a way to further develop expression (6.3). Thus, we now have that

$$f(x) = E[f(x + δ)] = E\left[f(x) + \frac{df(x)}{dx}δ + \frac{1}{2}\frac{d^2f(x)}{dx^2}δ^2\right]. \quad (6.4)$$

By using the rules for expectations introduced in Background 3:1, we can develop the right-hand expression as follows:

$$E\left[f(x) + \frac{df(x)}{dx}δ + \frac{1}{2}\frac{d^2f(x)}{dx^2}δ^2\right]$$

$$= E[f(x)] + E\left[\frac{df(x)}{dx}δ\right] + E\left[\frac{1}{2}\frac{d^2f(x)}{dx^2}δ^2\right]$$

$$= f(x) + \frac{df(x)}{dx}E[δ] + \frac{1}{2}\frac{d^2f(x)}{dx^2}E[δ^2].$$

(Here, $f(x)$, $\frac{df(x)}{dx}$ and $\frac{d^2f(x)}{dx^2}$ behave as constants; their expectations are therefore equal to these constants themselves.) When we take this back to (6.4), we see that we now know that

$$f(x) = f(x) + \frac{df(x)}{dx}E[δ] + \frac{1}{2}\frac{d^2f(x)}{dx^2}E[δ^2],$$

which can be simplified to

$$\frac{df(x)}{dx}E[\delta] + \frac{1}{2}\frac{d^2f(x)}{dx^2}E[\delta^2] = 0. \tag{6.5}$$

Now we need to be more specific about the δ values. However, we see that we do not need to study them 'individually'; it is sufficient that we, in simple ways, can express their summary statistics, $E[\delta]$ and $E[\delta^2]$.

Characterizing the δ values

For our further study of the δ values, we need to go back to our initial biological assumptions.

We have a diploid population of size N in which allele A_1 has frequency x at the start of a generation. The allele is favoured by selection of strength s; by using result (1) in Background 2:3, we know that its frequency in the gametic output of the considered generation therefore will be

$$x' = \frac{x(1 - sy/2)}{1 - sy},$$

where $y = 1 - x$. (Thus, x and y here take on the roles of p and q in Background 2:3.) For small values on s, we can simplify this as follows:

$$x' = \frac{x(1 - sy/2)}{1 - sy} = \frac{x - sxy + sxy/2}{1 - sy}$$

$$= x + \frac{sxy/2}{1 - sy} \approx x + sxy/2 = x + sx(1 - x)/2.$$

If the $2N$ gene copies that will make up the next generation are drawn at random from this gametic pool, then the number of A_1 copies – which we denote by N_{A_1}, an integer that may range from 0 to $2N$ – will follow the binomial distribution $Bin(2N, x + sx(1 - x)/2)$ (see Background 4:2). We are, however, not primarily interested in the distribution of the actual number of A_1 alleles in the new generation, but rather in their relative frequency. And, particularly, how this frequency deviates from the allele frequency in the generation before – thus, we are interested in the distribution of the δ values. This distribution we can write as $\frac{N_{A_1}}{2N} - x$, where for our present purpose x and N are constants and N_{A_1}

is the binomially distributed stochastic variable with parameters $2N$ and $x + sx(1 - x)/2$.

The rules for how to calculate the expectation and the variance for such a composite distribution are simple and are described in Background 3:1, while the basic properties of the binomial distribution are outlined in Background 4:2. With these tools, we are prepared for the final steps in our assault on expression (6.2). We start with the expectation of the δ values:

$$E[\delta] = E\left[\frac{N_{A_1}}{2N} - x\right] = \frac{E[N_{A_1}]}{2N} - x$$

$$= \frac{[x + sx(1 - x)/2] \cdot 2N}{2N} - x = sx(1 - x)/2. \qquad (6.6)$$

This result gives us no surprise, since it just tells us the expected increase in the frequency of A_1 between two generations, which we knew already.

When it comes to the more tricky question of finding a nice expression for $E[\delta^2]$, we start by remembering the simple rule for variances (see Background 3:1):

$$V[\delta] = E[\delta^2] - E[\delta]^2.$$

From result (6.6), it turns out that $E[\delta]^2$ must be of size s^2; since s is assumed to be small, we choose to ignore this term. Thus, we have:

$$E[\delta^2] \approx V[\delta].$$

We can now use our knowledge of how to calculate the variance of composite distributions plus how to calculate the variance of a binomial distribution:

$$V[\delta] = V\left[\frac{N_{A_1}}{2N} - x\right] = \frac{V[N_{A_1}]}{4N^2}$$

$$= [x + sx(1 - x)/2] \cdot [1 - x - sx(1 - x)/2] \cdot 2N/4N^2$$

$$= x(1 - x)/2N + \dots$$

The unfinished expression continues with terms all containing the factor s/N. Being the product of a small value with the inverse of a large number, we assume that all expressions with this factor are so small that they can be ignored.

We have now found that

$$E[\delta^2] = V[\delta] \approx \frac{x(1-x)}{2N},$$

and we are thus ready to gather all our partial results together.

Putting the parts together

Let us summarize. Our original equation (6.2),

$$f(x) = \sum_{\delta} p(x+\delta) \cdot f(x+\delta),$$

we reworked with the aid of a Taylor expansion and some approximations to obtain (6.5),

$$\frac{df(x)}{dx}E[\delta] + \frac{1}{2}\frac{d^2f(x)}{dx^2}E[\delta^2] = 0.$$

Now that we know more about the distribution of δ values, we can write this equation as

$$\frac{df(x)}{dx}sx(1-x)/2 + \frac{1}{2}\frac{d^2f(x)}{dx^2}\frac{x(1-x)}{2N} = 0.$$

For all values on x strictly greater than 0 and strictly less than 1, this equals

$$2Ns\frac{df(x)}{dx} + \frac{d^2f(x)}{dx^2} = 0,$$

and we have thereby reached our first goal. We have shown that the function describing the fixation probability of allele A_1 must fulfil the identity we originally gave as (6.1).

Before we proceed to solve the differential equation, let us reflect on the assumptions, simplifications and approximations that we have made in our struggle to reach the result. Some of these can be questioned – particularly the move we did when we ignored a number of small terms in the Taylor expansion. If the number of ignored terms is large, then maybe the *sum* of these very small terms is not so very small anymore, and they should therefore perhaps not have been so easily discarded?

Problematic also is our fundamental assumption that there should exist a nice continuous function giving the fixation probability of an allele with frequency x. This is so because our assumption that the number of gene copies in the population is $2N$ is central to the whole derivation. The allele frequency, x, is therefore not distributed over all the real numbers, but is limited to the numbers $\frac{N_{A_1}}{2N}$, where N_{A_1} is an integer between 0 and $2N$. In particular, it is strange that we assume continuity for f close to 0 and 1, when, for example, the smallest possible allele frequency is $1/(2N)$.

Instead of responding to these correct but not very worrying qualms with even deeper mathematical analyses – which can be done, but would not be trivial – it is nowadays much easier and more reasonable to refer to the excellent fit one gets between the theoretical results coming out of expression (6.1) and the results one obtains from computer simulations of the assumed process. Thus, after this moment of reflection, we leave the finer details of the derivation we have made to the experts and continue with our analysis.

Solving the equation

Expression (6.1) is a differential equation containing both the first and the second derivative of function $f(x)$, and it is therefore said to be of second degree. It is, however, so simple that we can find its solution by the use of the same tools as we presented in Background 1:2 and used in the first derivation.

Let

$$g(x) = \frac{df(x)}{dx},$$

and thus

$$\frac{dg(x)}{dx} = \frac{d^2f(x)}{dx^2}.$$

With this rewriting, equation (6.1) becomes

$$2Nsg(x) + \frac{dg(x)}{dx} = 0,$$

or

$$\frac{dg(x)}{dx} = -2Nsg(x).$$

In Background 1:2, we found the general solution to the differential equation $\frac{dy}{dx} = b - ay$. This result can be used here, and thus we find that

$$g(x) = 0 - Ce^{-2Nsx} = -Ce^{-2Nsx},$$

where C is any real number.

We now return to our definition of $g(x)$ with the knowledge that

$$\frac{df(x)}{dx} = Ce^{-2Nsx}, \tag{6.7}$$

where again C is any real number (the sign of C doesn't matter, of course).

From this follows the full set of solution

$$f(x) = C\left(-\frac{1}{2Ns}\right)e^{-2Nsx} + C^*, \tag{6.8}$$

where C and C^* are any real numbers.

We are not interested in all equations (6.8) that solve (6.1), but only in the specific one that satisfies the two boundary conditions $f(0) = 0$ and $f(1) = 1$. The first condition gives that

$$-\frac{C}{2Ns} + C^* = 0,$$

which implies that

$$C^* = \frac{C}{2Ns},$$

and the second that

$$C^* = 1 + \frac{C}{2Ns}e^{-2Ns}.$$

From this follows that

$$C = \frac{2Ns}{1 - e^{-2Ns}}$$

and

$$C^* = \frac{1}{1 - e^{-2Ns}}.$$

When these values for C and C^* are inserted into (6.8) and some rearrangements are made, we reach our ultimate goal:

$$f(x) = \frac{1 - e^{-2Nsx}}{1 - e^{-2Ns}}. \tag{6.9}$$

This is the major result of our derivation. We have now found the function for the probability that a positive mutation with current frequency x goes to fixation, and it turns out to critically depend on the product Ns. As demanded, $f(0) = 0$ and $f(1) = 1$, and in between these two points the function is strictly monotonically increasing with x. In fact, it raises steeply from 0 towards values close to 1 even for rather small x values, as long as Ns is allowed to be of reasonable size.

All the founders of population genetics used diffusion-like methods to tackle questions about genetic drift, but this clear and beautiful result was first explicated in the present form by Kimura (1962). Examples of function $f(x)$ as given by (6.9) are illustrated in Figure 6.1 for three different values of Ns.

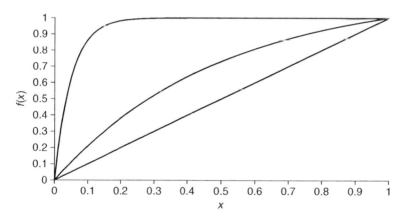

Figure 6.1 The probability, $f(x)$, for a selectively favoured allele to go to fixation as a function of its initial frequency, x (see expression (6.9)). The three curves are based on Ns, the product of the population size and the selection value, being equal to 0, 1 and 10, starting from below.

Derivation 6

The fate of a new single mutation

A case worth studying in detail concerns the fate of a newly arisen mutation. In a population of N diploid individuals, a new mutation would have starting frequency $x = 1/(2N)$. If this value is inserted into (6.9), we get

$$f\left(\frac{1}{2N}\right) = \frac{1 - e^{-s}}{1 - e^{-2Ns}}, \tag{6.10}$$

which for $2Ns$ greater than, say, three can be approximated with

$$f\left(\frac{1}{2N}\right) \approx \frac{1 - e^{-s}}{1 - 0} = 1 - e^{-s}, \tag{6.11a}$$

which is close to

$$f\left(\frac{1}{2N}\right) \approx 1 - (1 - s) = s. \tag{6.11b}$$

In the last step, we have used the standard approximation that $e^x \approx 1 + x$ when x is numerically small (see Background 2:1).

Thus, the probability for a new positive mutation to go to fixation in a large population is approximately equal to its selective advantage. This simple and useful result was first explicitly given by Haldane (1927).

Here, the reader should remember that we have assumed the relative fitness values of the genotypes to be 1, $1 - s/2$ and $1 - s$. When analysing the same situation, other authors may write the fitness values as 1, $1 - s$ and $1 - 2s$, or as $1 + 2s$, $1 + s$ and 1 (which, for small values on s, is exactly the same – see Question 3 in Derivation 2). In either case, result (6.11b) becomes $2s$ instead.

Summing up

Not all beneficial mutations will become fixed in a population, since all such mutations run a risk of being lost due to pure chance. The probability that a selected allele goes to fixation is a function of its current frequency and of $2Ns$ (i.e. twice the product of the population size and the allele's selective advantage). The fixation probability of a new, unique, positive mutation is closely approximated by its selective advantage and does not depend on the size of the population in which it appears.

Interpretations, extensions and comments

Drift and selection

The results we have reached open up for many questions to discuss. We start by giving some additional results about the fixation of positive mutations in finite populations. Then we extend our calculations by letting s, the positive parameter that measures the strength of selection, become negligible or even negative – and we see that the derivations we have done are still informative.

The probability of fixation of neutral mutations is then reconsidered using a completely different proof; an important result about the rate of turnover of neutral genetic material follows directly from it. Finally, we link our results to the discussion on variation and coalescence that we ended the preceding chapter with, and study how the variation at a locus is affected by selection at loci close by. Throughout this discussion section, we shift between insights about the processes of evolution in general and results of importance for the analysis of genetic data.

More on positive mutations going to fixation

From further use of diffusion arguments and other mathematical techniques, it is possible to derive some additional interesting results concerning the fixation of advantageous mutations. We give them without formal proofs (for such, see *e.g.* Crow and Kimura, 1970); they can also be confirmed and extended with suitable computer simulations.

The 'danger' a positive mutation runs is that it will lose out in the Mendelian lottery, when gene copies are drawn to form new generations, and be lost by pure chance (bad luck!). This risk is high only as long as the mutant allele remains very rare. As soon as a positive mutation has reached a nontrivial frequency and is present in more than a handful of copies, its chance of going to fixation becomes substantial.

This implies that if a mutation with a distinct fitness advantage has survived, let us say, ten generations after its introduction, then it is likely that it is present in so many copies that it will not subsequently be lost by genetic drift. Thus, it is primarily in the very earliest stages of a positive mutation's history that it may become lost due to genetic drift.

In our main derivation, we assume that the positive mutation acts in an additive way, so that the fitness of the heterozygotes is intermediary to the fitnesses of the two kinds of homozygotes. If instead the fitness regime is like 1, $1 - hs$ and $1 - s$ for genotypes A_1A_1, A_1A_2 and A_2A_2 (see Background 2:3), where h is the measure of a dominance effect, then the probability of fixation is actually already known to us, at least in principle and for values on h that are not close to 1.

This follows from the fact that new mutations are threatened with loss primarily in their initial stages, when they are still rare, and that from Hardy–Weinberg considerations we know that rare alleles are primarily found in heterozygotes. Thus, we can rewrite, and reinterpret, our earlier result (6.11b) that the probability of fixation of a new mutation is s as $2 \cdot (s/2)$, where $s/2$ is the difference in fitness between the rare heterozygotes and the common homozygotes. With this logic, we see that under the fitness scheme 1, $1 - hs$, $1 - s$, where the difference in fitness between the rare heterozygotes and the common homozygotes is $(1 - hs) - (1 - s)$ $= s(1 - h)$, the probability of fixation can be approximated by:

$$f\left(\frac{1}{2N}\right) \approx 2 \cdot s(1 - h) = 2(1 - h)s.$$

The negative influence of recessivity of selectively positive new mutations is seen from the last expression, in that the probability of fixation of a new mutation decreases with increasing values on h. When the positive mutation becomes strictly recessive ($h = 1$), our earlier assumptions are not valid anymore; instead, according to Kimura (1962), the following approximate result holds for the probability of fixation of such mutations:

$$f\left(\frac{1}{2N}\right) \approx \sqrt{\frac{s}{\pi N}}.$$

A simple numerical example shows that for $s = 1\%$, the probability of fixation of a new semi-dominant mutation ($h = 1/2$) is 1%, independent of population size (see expression (6.11b)), while for a strictly recessive mutation in a population of size $N = 10\,000$, it is 0.056%.

In simple population genetics models with no drift, it can be shown that a positive mutation that has reached a nontrivial frequency in the population will quite rapidly become very common. The number of generations it takes for the mutation to go from a small to a large frequency is directly related to the inverse of the selective advantage. Thus, a mutation that is twice as good as another will only take half as long to go from rare to common (a result first presented by Haldane in 1924; for some numerical values, see *e.g.* Hedrick, 2011a).

It turns out that this deterministic result also holds reasonably well for limited populations in which genetic drift has to be taken into account. In the model that we have used here, a new positive mutation that ultimately goes to fixation in a population of 10 000 diploid individuals takes on average 2326 ± 29 generations to do so if $s = 0.01$ (*i.e.* for an advantage of 1%). For $s = 0.02$, the expected time is 1285 ± 13 generations. (These values have been obtained from computer simulations with 197 and 213 runs to fixation, respectively.)

The probability of fixation of different kinds of mutations

It came as a shock to the early population geneticists when they realized how likely it is that a truly favourable mutation does not go to fixation but instead ends up being lost. As good Darwinists, they put a high value on positive mutations and saw it as a waste if these mutations were not recruited to advance the adaptation of the population. And, as we have seen, the probability for this kind of loss is certainly substantial: according to Haldane's result (6.11), a new mutation giving heterozygotes a 1% and homozygotes a 2% fitness advantage (enormous values in most ecological situations, if stably sustained) will with 98% probability be lost without any evolutionary trace.

This makes it reasonable to study in more detail different types of mutations and the frequencies with which they go to fixation. Here, our careful description of how to obtain the fixation function $f(x)$, as given by (6.9), becomes useful. In its derivation, we assumed that the population size, N, is large and that the advantage of the new allele, s, is small and positive. If one goes back and looks carefully at the different steps in the derivation, it is seen that the assumption of s being positive is only used to motivate the model as a description of the behaviour of an advantageous mutation. In fact, this assumption about the positivity of s never plays any important role in the calculations. The whole derivation works equally

Derivation 6

well if s is a negative number, except that (6.10) is then better written

$$f\left(\frac{1}{2N}\right) = \frac{e^{-s} - 1}{e^{-2Ns} - 1} = \frac{e^{|s|} - 1}{e^{2N|s|} - 1},$$ (6.12a)

and the subsequent approximation for, say, $2N|s| > 3$ instead becomes

$$f\left(\frac{1}{2N}\right) = \frac{e^{|s|} - 1}{e^{2N|s|} - 1} \approx \frac{e^{|s|} - 1}{e^{2N|s|}} = (e^{|s|} - 1)e^{-2N|s|}$$

$$\approx [(1 + |s|) - 1]e^{-2N|s|} = |s|e^{-2N|s|}.$$ (6.12b)

But what happens to our derivation if we let s be exactly equal to 0; that is, if the new mutant we consider is *exactly* selectively neutral?

Again, the derivation of the diffusion equation (6.1) may be left unchanged – which increases our conviction that the equation gives a well-founded description of the interplay between drift and selection in our model. However, when it comes to *solving* the diffusion equation, matters become different and, actually, much simpler, since with $s = 0$, the differential equation (6.1) reduces to only one term:

$$\frac{d^2 f(x)}{dx^2} x(1 - x) = 0.$$

In the interior of the interval from 0 to 1, this means that

$$\frac{d^2 f(x)}{dx^2} = 0,$$

which is solved by any straight line $ax + b$. Therefore, given the boundary conditions $f(0) = 0$ and $f(1) = 1$, the single relevant solution becomes

$$f(x) = x.$$ (6.13)

We have now reached another important result (to which we soon will return, with a very different proof): the probability of fixation of a neutral allele with frequency x is exactly x. We have also seen how useful a carefully outlined derivation may be, in that it can easily be returned to and reanalysed for interesting aspects not immediately realized. (The scientific importance of being able to return to earlier proofs is stressed by the Hungarian-English mathematical philosopher Imre Lakatos in *Proofs and Refutations* (1976); a highly stimulating book, although difficult in parts.)

The probabilities of fixation of new alleles associated with fitness values ranging from negative over 0 to positive have now been established for our model. The only restrictions are that s should be numerically small and that the population size should be reasonably large (and these demands are not particularly limiting, as can be shown with computer simulations; our results function well as long as s is numerically smaller than 5% and N is larger than 100). The relationship we have derived between the strength of selection, s, and the probability of fixation of a new mutation can be numerically illustrated as follows:

$s =$	$-\dfrac{8}{2N}$	$-\dfrac{4}{2N}$	0	$\dfrac{4}{2N}$	$\dfrac{8}{2N}$
$f\left(\dfrac{1}{2N}\right) =$	$\dfrac{0.0027}{2N}$	$\dfrac{0.075}{2N}$	$\dfrac{1}{2N}$	$\dfrac{4.07}{2N}$	$\dfrac{8.00}{2N}$

Here, all values are expressed in terms of $\frac{1}{2N}$, which is a natural measure since it is the lowest allele frequency there can be; it is also the probability that a single new neutral mutation will go to fixation.

Within the illustrated range of s and N values, it is seen that a single neutral mutation has a probability of fixation equal to its frequency, and that a positive mutation has a probability of fixation approximately equal to its selective advantage (all according to the simplified results for $2Ns > 3$ described earlier). When s is slightly negative, the mutation has a distinctly lower probability of fixation that a neutral mutation, but its probability of fixation is not negligible. More clearly negative mutations are not likely ever to become fixed due to genetic drift alone.

The greatest surprise in these results is, probably, that slightly deleterious mutations have a non-negligible chance of going to fixation. They are certainly less likely than effectively neutral mutations to do so, but as long as their fitness disadvantage has a numerical value not much greater than the inverse of the population size, their possibility of fixation must be taken into account (the preceding table describes what happens for twice and four times this value). The Japanese geneticist Tomoko Ohta (1973) was the first to stress the importance in evolution of such deleterious but nearly neutral mutations.

An alternative proof of the probability of fixation under neutrality

Before we continue our discussion of the rates with which different types of mutations go to fixation, let us give an alternative proof of the result we reached in (6.13), that the probability of fixation for a neutral mutation equals its current frequency. The following argument has the advantage of being very general and without any limiting restrictions beyond the absence of new mutations.

Assume that a neutral allele, A_n, has frequency x in a finite population (whose size may vary between generations). This means that if we draw a gene copy from the population, then the probability of it being of allelic state A_n is x. For this type of genetic variation, we know that the Law of Constancy of Allele Frequencies holds (as described in the Introduction). Thus, the expected frequency of finding A_n if we draw a gene copy from the next generation is also x. And so it will be for all future generations.

This leaves us with what first may appear as a paradox: the probability that a drawn gene copy is A_n continues to be x for all subsequent generations, even when we know that a population of this type with no mutations gradually will lose its variation and become monomorphic with time (as discussed in the beginning of Derivation 5).

There is, however, a simple solution to this conundrum: either the population becomes fixed for A_n with probability x, or the allele is lost with probability $1 - x$. In this way, there will be a complete loss of genetic variation in the population at the same time as the probability to draw a gene copy of type A_n remains x for ever. The Law of Constancy of Allele Frequencies is thereby satisfied, even when the population we study becomes devoid of any genetic variation.

With this argument, we have, once more, showed that the probability of fixation of a neutral allele equals its current frequency, but we have done so without any approximations or assumptions about the size of the population.

The rate of fixation of neutral mutations and evolutionary clocks

Two additional results about the fixation of neutral mutations are important to notice.

The first concerns the time it takes for a new neutral mutation to go to fixation in a population, assuming that it does so. The expected time for such a fixation process can be shown to be $4N$ generations (in populations where N is not small). We have, actually, been close to obtaining this result already, when in Derivation 4 we found that the mean time to coalescence of n gene copies tends towards $4N$ when n increases (see expression (4.4)). In that situation, our result was reached assuming $n < N$, but removing this constraint does not affect the result in any significant way (Hein *et al.*, 2005). We have here yet another interesting example of how population genetics can treat questions by going either back in time or forward in time. Thus, the question about the time to fixation of a single gene copy turns out to be identical to the question of the time to ultimate coalescence of all gene copies in a population.

This time is normally much, much longer than the time it takes for new *positive* mutations to go to fixation, as seen in the example with a diploid population of size 10 000. Here, a mutation with a fitness advantage of 1% takes about 2300 generations to go to fixation, while a selectively neutral mutation takes on average 40 000 generations to become fixed.

The second property of neutral fixation that is relevant to discuss concerns its rate under recurrent mutation. In a small population, neutral mutations occur rarely, but each mutation has a relatively high chance of becoming fixed. In a large population, many new neutral mutations occur every generation, but each mutation has a fairly small probability of going to fixation. It turns out that these two processes exactly compensate for each other, leading to a turnover rate of mutations in limited populations that is independent of the population size, given a fixed mutation rate.

In a more technical form, the argument goes like this: In a diploid population of size N, with mutation rate μ, the expected number of new mutations is in every generation $2N\mu$. Each one of these mutations has a probability of $1/(2N)$ to go to fixation, which makes the product of these two numbers exactly equal to μ. Thus, the rate of turnover of neutral genetic material in finite populations is only determined by the rate with which new such mutations occur, and not by the size or structure of the population in which the material exists.

This fundamental result was pointed out by Kimura (1968) when he first presented his idea that most variation detectable at the molecular level is selectively neutral. This constituted a revolution for population genetics. With his book *The Neutral Theory of Molecular Evolution* (1983), Kimura became the leader in shifting theoretical population genetics from analyses of how the rules of genetics interact with evolutionary forces to shape organisms and their adaptation to the natural environment, towards analyses of how the genetic material in itself carries information about what processes have affected it in its evolutionary past.

Among the many different aspects of neutral evolution, let us here point only to the key importance of the result just obtained for any notion of an 'evolutionary clock' – the suggestion that there is a chronologically steady turnover of parts in the genome that can be used to time past evolutionary events, in particular the divergence of taxa. Every theoretical support of this idea is based on the argument about the rate of fixation of neutral mutations we have just worked through.

Recurrent mutations and near neutrality

We will now continue to discuss the rates at which different classes of mutations go to fixation, seen from a general perspective.

Even if we have stressed that positive mutations may not necessarily go to fixation, it is clear that if such mutations recur reasonably frequently, then they will sooner or later become fixed. As to the rate of fixation of (effectively) neutral mutations, it is given by their mutation rate alone. When it comes to slightly deleterious mutations, their probability of fixation is low, but they are on the other hand likely to occur at a relatively high frequency; certainly much higher than the frequency with which positive mutations occur.

This makes it rational to believe that most mutations that go from rare to common in populations over time have small, very small or almost no effect at all on fitness. The distinctly positive mutations may be of great evolutionary importance, but they will not dominate in the gradual change of the genetic material. The distinctly negative mutations (a very large class) we can ignore in this context; they will normally never take part in the genetic turnover. This leaves us with the class of effectively neutral and nearly neutral mutations. Any differentiation between these two classes will normally be impossible in concrete situations; only rarely can fitness measurements of such precision be made.

An important insight is, however, gained from these general considerations. From our earlier calculations, we learnt that mutations with fitness effects not much greater than $1/N$ will behave as nearly neutral. This implies that the classification of a mutation as nearly neutral must partly be seen as a population phenomenon and not solely as a description of the mutation's intrinsic phenotypic quality. A mutation that is nearly neutral in a small population may well behave as deleterious in a large population.

We have here another reason for not taking the key result (5.2b) in the last chapter, saying that the degree of variation at a locus is given by the expression $H \approx \frac{4N\mu}{1+4N\mu}$, too rigorously. The rate with which new mutations occur, μ, appears there as a constant, but it can – as we just have seen – be considered as a function of the population size, N, itself. When this is done, the expected rate with which neutral mutations occur (be they strictly neutral, effectively neutral or nearly neutral) decreases with the size of the population. Thus, when it comes to the analysis of actual DNA sequences, it is normally impossible to distinguish between such genetic variation that is under selection and such variation that is neutral and evolves solely according to stochastic rules.

To end this chapter, we extend the final point by noting that our analysis so far has been restricted to the variation that exists at a single locus seen, in some sense, as being completely independent of the rest of the genome. What happens to neutral genetic variation that is linked to a locus evolving under natural selection?

Selective sweeps, the hitch-hiking effect and background selection

Let us consider a new positive mutation that will increase in frequency by natural selection until it becomes fixed in the population. When the mutation occurs, it happens at a site on a chromosome and starts in absolute linkage disequilibria with all genetic variants along this chromosome. When the mutation then increases in frequency, so does the chromosome on which it sits, plus all the associated genetic variants. The associations between the selected mutation and the genetic markers that are not closely linked will soon disappear due to recombination, but the associations with closely linked markers will remain for a long time (see Figure 3.1!) – one way to envisage this situation is to see natural selection as dragging not just the mutation but a whole chromosome region towards fixation. Correspondingly, one often talks about the linked marker alleles, with no causal relationship to the positive effect of the mutation, as hitch-hiking in evolution.

Variants somewhat further along the chromosome will normally not go all the way to fixation, but their frequencies will still be affected. All this implies that the mean time to coalescence for neutral sites close to the positively selected locus will be smaller than otherwise expected. Or, in other words: the level of variation at these marker sites will be smaller than elsewhere in the genome.

Thus, in an interval around a gene where a positively selected allelic fixation has just occurred, a reduced level of neutral variation will be seen, an effect first described in detail by Maynard Smith and Haigh in 1974. This loss of variation can be regarded as a 'selective sweep' – a local cleaning away of neutral variation. In genomic studies of genetic variability, it is therefore often of great interest to find a chromosomal stretch where the genetic variation is significantly reduced, since this may indicate that the region contains a site where a mutation recently has been favoured by selection. Interesting examples of this have, for example, been seen in the vicinity of the human gene affecting lactose tolerance, where locally reduced haplotypic variation has been detected in Europeans (Bersaglieri et al., 2004), as well as in Maasai (Schlebusch et al., 2013) – two populations where lactase tolerance recently has been strongly selected for.

Have we here the explanation for the interesting phenomenon – mentioned briefly in the preceding chapter – that the level of neutral variation is, on average, lower close to active genes than elsewhere in the genome?

Probably not. Positive mutations that go to fixation reasonably fast and thereby sweep away variation from adjacent chromosomal regions are not so common that they may cause this general effect. Instead, what is called 'background selection' is its likely mechanism (Charlesworth *et al.*, 1993). A chromosomal region with important genes will be hit by deleterious mutations more often than a region without such genes, making the number of chromosomes from which future generations derive smaller in the first case than in the second. One way to express this effect is to say that the effective population size is smaller for gene-rich regions than for gene-sparse regions. This difference in population size translates into an expectation that there should be a lower frequency of neutral variation close to functional genes than elsewhere in the genome.

These discussions show that a certain precision is needed when the effect of selection on the genome is discussed. Assuming near or effective neutrality for a particular genetic variant in a specific situation and over a restricted timescale may be reasonable and sound, but claiming that the corresponding locus/gene has been unaffected by selection over a long evolutionary period is a much more questionable matter. Particularly in large populations, it may be that no position in the genome evolves without ever in one way or another being influenced by direct or indirect natural selection, if only ever so slightly.

In general, it should be remembered that selection and genetic drift do not function as alternative evolutionary principles. Genetic drift is always present, since biological populations are limited in size, while selection is a fact of life that constantly affects all genetic material, though with variable strength and via different mechanisms.

Background 6:1 Taylor expansion

Consider a continuous and infinitely differentiable function, $f(x)$. According to a key result in mathematical analysis, it follows that for a small positive or negative change, δ, we can write:

$$f(x + \delta) = f(x) + \frac{df(x)}{dx}\delta + \frac{1}{2}\frac{d^2f(x)}{dx^2}\delta^2 + \sum_{i=3}^{\infty} \frac{1}{i!}\frac{d^if(x)}{dx^i}\delta^i. \qquad (1)$$

(Since the prime sign, ', in population genetics normally is used to designate 'in the next generation', we are here careful not to let it denote derivation.)

This expression, called a Taylor expansion, is useful in many ways. For example, if the values of the function and its derivatives are known for a specific point, x, then the value of the function at any nearby point $x + \delta$ can be calculated. It may also be used, as is done in this chapter's main derivation, to approximate a complicated function by a polynomial which then can be suitably simplified. Thus, in our genetic derivation, it is posited that all terms containing δ^3 and higher powers in δ in the relevant polynomial are so small that they can be ignored.

It may be noticed that when we obtained the linear functions for perturbations around the equilibrium points in Derivation 2, we in fact performed the first step in a Taylor expansion.

Questions

1. What is the probability of fixation in a large population of a new mutation so favoured by selection that $s = 0.005$? What should the frequency of a selectively neutral mutation be for it to have the same probability of fixation?

2. Does a positive mutation with fitness value $s = 1\%$ and population frequency 1% have a higher chance of fixation than a mutation with $s = 0.5\%$ but population frequency 2%?

3. What is the probability that a new *deleterious* mutation goes to fixation in a diploid population of size 10 000 if its selective disadvantage is 10^{-5}, 10^{-4} or 10^{-3}?

4. After an abrupt environmental change, a very rare type of mutation becomes strongly advantageous ($s = 1\%$). Its rate of mutation is 10^{-7}, and when we start to study our population of constant size 5000, there is no copy of the mutation in it. Describe a likely scenario for the fixation of this mutation.

5. We have found the probability of fixation, $f(x)$, for a mutation with frequency x. Since – given a sufficiently long time – all genetic variants will either go to fixation or become lost, it is trivial to see that the probability of loss of an allele, $l(x)$, equals $1 - f(x)$. The function $l(x)$ can, however, also be deduced from first principles. By looking at the derivation of $f(x)$, sketch the corresponding derivation of $l(x)$! (This problem is actually very easy; it also illustrates, once again, how useful it is to be able to return to old proofs.)

Derivation 7

Nonhomogeneous populations

Until now, we have treated populations as homogeneous by assuming that random mating occurs in them, which means that all individuals in the population have equal chance of meeting and mating with one another. In reality, this is of course rarely the case — for some pairs, it is easier to meet and mate than for others. Another problem worth reflecting upon is how to grasp the structuration of individuals belonging to the same species in nature. Only seldom can one directly say where one population ends and another starts. Or whether a studied population is homogenous in its composition or consists of two or more partially differentiated subpopulations.

We treat the broad and relevant questions of population structuration by first describing a way to theoretically capture the phenomenon in its most distinct form: as a subdivision of the population into discrete parts, more or less genetically differentiated. The main derivation shows that such subdivision always leads to a perceived lack of heterozygotes. We then introduce a commonly used measure of how strong a given subpopulation differentiation is, F_{ST}. Together, these achievements provide an understanding of what genetic effects the nonhomogeneity of real populations leads to and how these effects can be studied and summarized.

Since we are primarily interested in how the lack of homogeneity of real populations affects the distribution of their *current* genetic variation, we perform the main part of our derivation without specifying population numbers. Later, we broaden the discussion, however, to describe more specific population models that can be used for studying how the

Understanding Population Genetics, First Edition. Torbjörn Säll and Bengt O. Bengtsson.
© 2017 John Wiley & Sons Ltd. Published 2017 by John Wiley & Sons Ltd.

subpopulations' limited size and/or geographical proximity affects their genetic differentiation over time.

We end the chapter with a statistical interpretation of the key measure F_{ST} and a historical note on the background to the theoretical approach used – which turns out to be most concrete and far from theoretical.

Analysis

Assumptions, notations and definitions

Assume a population that is divided into m subpopulations, $1, \ldots, m$, each of relative size c_k, so that $\sum_{k=1}^{m} c_k = 1$. The population and all the subpopulations are so large that chance effects can be ignored. We will be concerned with a locus, A, with alleles $A_1, A_2, \ldots$ and overall allele frequencies $p_1, p_2, \ldots$ The frequency of allele i in population k will be denoted $p_{i,k}$. The overall frequency of allele i is then the average over subpopulations: $p_i = \sum_{k=1}^{m} c_k p_{i,k} = \bar{p}_i$. We will mostly use the notation $\bar{p}_i$ rather than just p_i, since this makes explicit that the allele frequency p_i is an average (this particular notation is introduced in Background 3:1). We will also assume that there is local random mating within all the subpopulations, so that the relative frequency of homozygotes with genotype $A_i A_i$ in subpopulation k is $P_{ii,k} = p_{i,k}^2$, and the relative frequency of heterozygotes $A_i A_j$ in the same population is $P_{ij,k} = 2p_{i,k}p_{j,k}$. (This is where the assumption about the population sizes being large enters the derivation; if we considered small populations, then the sampling variance in the formation of diploids would have to be reckoned with.) The set-up that we now have presented was originally introduced by Wahlund (1928).

Our aim is to determine the genotype frequencies in the combined (overall, total, whole) population as a function of the distribution of the allele frequencies over the subpopulations. These genotype frequencies will be compared to what the frequencies would be in the absence of any subdivision; that is, with random mating in the whole population. Before we start the derivation proper, let us first remind ourselves of the usefulness of measure H.

A reminder: expected heterozygosity

Expected heterozygosity H (also called 'genetic diversity', and as such discussed in Derivations 3 and 5) will be used here as a convenient measure of genetic variation. The term was introduced earlier as the probability that two randomly drawn gene copies are different, and it retains the same meaning here. Thus, given a locus A, with alleles A_1, A_2, ... and allele frequencies p_1, p_2, ..., the expected heterozygosity for the locus is $H = 1 - \sum_i p_i^2$ (this follows from the fact that if two gene copies are not of the same allelic type, then they must be different). In the situation with only two alleles at the locus, this reduces to

$$H = 1 - p^2 - q^2 = 2pq,$$

where $p = p_1$ and $q = p_2 = 1 - p_1$. Thus, in random-mating populations, the heterozygosity H of a locus equals the frequency of heterozygotes in the diploid phase. However, this simple result does not hold when mating in the population is not random – as in the situation considered here, where random mating occurs only within the boundaries of the subpopulations.

Genotype frequencies in the combined population

Our primary task is to find the genotype frequencies when the population is considered in its entirety, even though it is subdivided, and we start by studying the homozygous genotypes. The frequency of genotype A_iA_i in the combined population, P_{ii}, is:

$$P_{ii} = \sum_{k=1}^{m} c_k P_{ii,k} = \sum_{k=1}^{m} c_k p_{i,k}^2 = \overline{p_i^2}. \tag{7.1}$$

To express this result in a more informative way, we remember from Background 4:2 that, in general,

$$V[p_i] = \overline{p_i^2} - \overline{p_i}^2,$$

and that therefore,

$$\overline{p_i^2} = \overline{p_i}^2 + V[p_i].$$

Here, $V[p_i]$ denotes the variance over the subpopulations of the allele frequency p_i. This result implies that we can rewrite expression (7.1) as

$$P_{ii} = \bar{p}_i^{\,2} + V[p_i]. \tag{7.2}$$

Thus, the overall frequency of a particular homozygous genotype is the square of the average of the frequency of this allele *plus* the variance over the subpopulations of this frequency. It is important to note that $\bar{p}_i^{\,2}$ would be the genotype frequency if the whole population mated completely at random ($V[p_i]$ would then automatically become 0). Since variances are positive, our result shows that subdivision of a population always leads to an overall relative increase in homozygotes for those alleles whose frequencies vary among subpopulations.

We now consider what happens to the heterozygotes. From the preceding paragraph, it is clear that in any subdivided population with genetic differentiation, there will in the total population always be what appears as a deficiency of heterozygotes. If all homozygous genotypes are overrepresented compared to their expected Hardy–Weinberg proportion, then the heterozygotes *must* be fewer than expected.

The phenomenon that the subdivision of a population into separate genetically differentiated reproductive parts always results in a relative increase in homozygotes and a relative lack of heterozygotes is known as the *Wahlund effect* (after Wahlund 1928).

It is interesting to study how this general conclusion applies when one considers a particular pair of alleles, say A_i and A_j, in detail. The overall frequency of the heterozygotes A_iA_j, P_{ij}, according to our assumptions, will be

$$P_{ij} = \sum_{k=1}^{m} c_k\, P_{ij,k} = \sum_{k=1}^{m} c_k 2p_{i,k}p_{j,k} = 2\sum_{k=1}^{m} c_k p_{i,k}p_{j,k} = 2\overline{p_i p_j}. \tag{7.3}$$

To proceed with our analysis, we now take advantage of the statistical notion of a *covariance* between two variables. Background 7:1 reminds us that

$$Cov\,[p_i, p_j] = \overline{p_i p_j} - \bar{p}_i \bar{p}_j,$$

which can be rewritten,

$$\overline{p_i p_j} = \overline{p}_i \overline{p}_j + Cov\,[p_i, p_j].$$

Accordingly, we can write (7.3) as

$$P_{ij} = 2\overline{p}_i \overline{p}_j + 2Cov\,[p_i, p_j]. \tag{7.4}$$

If the frequency of one allele at a locus goes up, then the frequencies of the other alleles at the locus will on average go down. The covariances between alleles in situations like the one outlined here will therefore on average be negative. Thus, in a subdivided population, the overall frequency of heterozygotes, P_{ij}, will normally be less than the frequency expected under random mating, $2\overline{p}_i \overline{p}_j$. One can, however, envisage situations with more than two alleles where not *every* heterozygote genotype at a locus is negatively affected by the population differentiation, even if the heterozygotes *in toto* are so (see Question 4 at the end of the chapter).

A special case: two alleles

We now take a closer look at the case when there are only two alleles at locus A, so that $p_{1,k} = p_k$ and $p_{2,k} = q_k = 1 - p_k$. The three genotype frequencies can then be expressed in the following way, using (7.2) and (7.4):

$$P_{11} = \overline{p}^2 + V[p],\, P_{12} = 2(\overline{p})(\overline{q}) + 2Cov\,[p, q] \text{ and } P_{22} = \overline{q}^2 + V[q].$$

However, since $q_k = 1 - p_k$, we are reminded by Background 7:1 that $V[q] = V[p]$. Furthermore, since $P_{11} + P_{12} + P_{22} = 1$ and $\overline{p} + \overline{q} = 1$, which implies that $(\overline{p} + \overline{q})^2 = 1$ and that $\overline{p}^2 + 2\overline{p}\overline{q} + \overline{q}^2 = 1$, we can write

$$1 = P_{11} + P_{12} + P_{22} = \overline{p}^2 + V[p] + 2\overline{p}\overline{q} + 2Cov\,[p, q] + \overline{q}^2 + V[q]$$

$$= \overline{p}^2 + 2(\overline{p})(\overline{q}) + \overline{q}^2 + V[p] + 2Cov\,[p, q] + V[q]$$

$$= 1 + 2V[p] + 2Cov\,[p, q].$$

It has thereby been shown that $Cov\,[p, q] = -V[p]$.

At a locus with only two alleles, it is, thus, always true that the covariance of the two alleles over the subpopulations is negative.

The three genotype frequencies can then be written

$$P_{11} = \bar{p}^2 + V[p], P_{12} = 2\bar{p}\bar{q} - 2V[p] \text{ and } P_{22} = \bar{q}^2 + V[p], \quad (7.5)$$

which is a common way to formulate the Wahlund effect for two alleles. The term $V[p]$ is generally called the *Wahlund variance*.

Subdivision resembles inbreeding

We have now seen that the effect of population subdivision with concomitant allele frequency differentiation is that the overall frequency of homozygotes is greater than the frequency expected under random mating. This makes subdivision resemble the effect of *inbreeding* in the sense of mating between close relatives (see the discussion to Derivation 4), even though subdivision is quite different from inbreeding in any strict biological sense. Remember that we have assumed the subpopulations are large and random-mating (and therefore that there are no matings between close relatives)!

Despite this difference between inbreeding and population differentiation, it is common to express (7.5) in the notation that traditionally is used for inbreeding *sensu stricto*. Recall that in a population with an average inbreeding coefficient F, the genotype frequencies for a locus with two alleles are as follows (see Background 2:1):

$$P_{11} = p^2 + Fpq, P_{12} = 2pq - 2Fpq \text{ and } P_{22} = q^2 + Fpq.$$

Thus, if we want to use F to express the deviation from Hardy–Weinberg proportions in the overall population that we just have derived, it follows from (7.5) that $F\bar{p}\bar{q} = V[p]$, which implies that

$$F = \frac{V[p]}{\bar{p}\bar{q}}.$$

This is a common way of quantifying the amount of genetic differentiation over subpopulations, due to Wright (1951). By tradition and convention, this special form of F is denoted F_{ST}, where S stand for the

Subpopulation and T for the Total (combined) population. Thus, we have that

$$F_{ST} = \frac{V[p]}{\overline{p}\,\overline{q}}. \tag{7.6}$$

F_{ST} takes the form of a variance divided by a normalizing factor, which makes it possible to compare the amount of genetic differentiation between loci with different allele frequencies. The expression $\overline{p}\,\overline{q}$ gives the highest possible value for the variance of p over subpopulations, which occurs when a fraction $\overline{p}$ of them are fixed for the A_1 allele while a fraction q are fixed for the allelic alternative A_2. The F_{ST} value can thus be seen as a measure of how big the differentiation between the subpopulations is relative to its maximal possible value. F_{ST} is by convention defined as 0 when the same allele is fixed in all subpopulations (when expression (7.6) formally becomes 0/0).

The reader is reminded of what was written earlier: the similarity between the effects of population subdivision and inbreeding that made us introduce the measure F_{ST} does not imply that these biological processes should be too closely identified. A high F_{ST} value may, for example, be found in situations with very large, perfectly random-mating subpopulations that are highly genetically variable and show no particular signs of being inbred.

Another way to express F_{ST} is in terms of heterozygosity values, taking advantage of how heterozygosity can be regarded as a measure of genetic variability. In order to do this, we first have to rewrite $V[p]$. We start with the result that was utilized to derive (7.2),

$$V[p] = \overline{p^2} - \overline{p}^2.$$

This result, we can develop and rewrite as follows:

$$V[p] = \overline{p^2} - \overline{p} + \overline{p} - \overline{p}^2 = (\overline{p} - \overline{p}^2) - (\overline{p} - \overline{p^2}) = \overline{p}(1 - \overline{p}) - (\overline{p} - \overline{p^2})$$

$$= \overline{p}\,\overline{q} - (\overline{p} - \overline{p^2}). \tag{7.7}$$

The first part of this expression is easily interpreted, since the expected heterozygosity in the combined population, which we will denote by H_T, is $2\bar{p}\bar{q}$. Thus, we have that

$$\bar{p}\bar{q} = \frac{1}{2}H_T, \tag{7.8}$$

where the index T stands for the total population. For the second part of the expression, recall from the beginning of this derivation that

$$\bar{p} = \sum_{k=1}^{m} c_k p_k$$

and from (7.1) that

$$\overline{p^2} = \sum_{k=1}^{m} c_k p_k^{\,2}.$$

Thus,

$$(\bar{p} - \overline{p^2}) = \sum_{k=1}^{m} c_k p_k - \sum_{k=1}^{m} c_k p_k^{\,2} = \sum_{k=1}^{m} c_k(p_k - p_k^{\,2})$$

$$= \sum_{k=1}^{m} c_k p_k(1 - p_k).$$

Also, this part can be given a simple interpretation, since the mean of the heterozygosity in the different subpopulations is obviously $\sum_{k=1}^{m} c_k 2 p_k(1 - p_k)$, which we will denote by $\overline{H}_S$, where the index S stands for subpopulation. Thus, we have that:

$$(\bar{p} - \overline{p^2}) = \sum_{k=1}^{m} c_k p_k(1 - p_k) = \frac{1}{2} \sum_{k=1}^{m} c_k 2 p_k(1 - p_k) = \frac{1}{2}\overline{H}_S. \tag{7.9}$$

If the results of (7.7), (7.8) and (7.9) are inserted into (7.6), then we get the attractive formula

$$F_{ST} = \frac{H_T - \overline{H}_S}{H_T}. \tag{7.10}$$

According to this expression, F_{ST} can be seen as a measure of that proportion of the genetic variation in the total population that is not due to variation *within* the different subpopulations but due to variation *between* them.

Summing up

If a large population is divided into a number of subpopulations, more or less genetically differentiated, then the total population will show a lack of heterozygotes compared to the expected Hardy–Weinberg proportions, even if there is local random mating within all the subpopulations. This lack of heterozygotes can be associated with an inbreeding measure, F_{ST}, which tells how much of the genetic variation in the total population is due to the differentiation between subpopulations. The magnitude of the F_{ST} value may, for example, be critical in a conservation biological situation, where the question is whether and how to manage a fragmented species.

Interpretations, extensions and comments

Genes and balls, populations and urns

All living organisms carry genes, which are stable entities that only rarely mutate into new types. This makes the situation very similar to the one generally used in statistics, when probability theory is taught with the aid of (imaginary) balls in urns. Indeed, one may see population genetics as being the study of the behaviour of such 'balls with different colours' (*i.e.* genes of different allelic types) in one or more 'urns' (*i.e.* populations). Sometimes, the underlying assumptions make the biology involved very explicit – this is, for example, the case in the next derivation, where the genes studied belong to sibs produced by diploid parents. At other times, very few biological assumptions enter the considerations. This, then, makes the results applicable to a broad range of organisms. Such is the case here when it comes to using F_{ST} as a measure of differentiation. Mathematically, the analysis we performed in the derivation is very close to describing the distribution of different-coloured balls over separate urns. Results such as (7.10) can, thus, be seen as following from an Analysis of Variance, separating the variation in a large population into within- and between-subpopulation components. This generality also means that many of the ideas outlined here can be applied to all kinds of organisms, may they be haploid or diploid, prokaryotes or eukaryotes. Even viruses can be analysed with these tools.

The situation becomes, however, somewhat special when the analysis is based on samples of diploids. This is because it is then possible to directly see whether the pairs of gene copies – as diploid organisms can be considered to be – have indeed been produced by a random process or not. Any bias detected in a sample of diploids, showing for example that there are 'too few heterozygotes', is an indication that these diploids were produced by some other process than a true random union of gametes.

(The occasional deviation from Hardy–Weinberg proportions in analysed population samples of out-breeding species is, however, more often due to problems with genotype recognition than to any interesting population processes – in our experience. This illustrates how important it is even for theoretical population geneticists to care about how available data have been produced.)

Population stratification and linkage disequilibria

Subpopulation differentiation can be detected, not only by deviations from Hardy–Weinberg proportions, but also by the presence of significant linkage disequilibria. We will not here go into the details of such analyses; let us just illustrate the effect with a simple example which we now take from a haploid organism.

Assume that in a sample, four-fifths of the individuals come from a subpopulation with allele frequencies $p(A_1) = 0.80$, $p(A_2) = 0.20$, $p(B_1) = 0.90$ and $p(B_2) = 0.10$ and no linkage disequilibrium between the loci, while the rest come from another subpopulation with allele frequencies $p(A_1) = 0.30$, $p(A_2) = 0.70$, $p(B_1) = 0.50$ and $p(B_2) = 0.50$ and, again, no linkage disequilibrium. In the combined sample, the frequencies of the four haplotypes will be $p(A_1B_1) = 0.606$, $p(A_1B_2) = 0.094$, $p(A_2B_1) = 0.214$ and $p(A_2B_2) = 0.086$, and now there is a linkage disequilibrium between the loci of size $D = 0.032$. This result is in some sense self-evident, since in the outlined situation the mixing of the two gene pools produces a positive correlation between carrying allele A_1 and carrying allele B_1, and it is this correlation that is detected as a linkage disequilibrium. (An illustration of how such disequilibria caused by mixing can be used to analyse a population's background history is illustrated by Question 5 in Derivation 3.) Observe, however, that we here have a linkage disequilibrium between two loci that may be completely unlinked; concluding from the presence of their correlation that they must be physically close on the same chromosome would, of course, be erroneous.

Since all association studies build on detecting linkage disequilibria between DNA markers closely linked to loci involved in the causation of the trait, this numerical example shows how important it is that all possible sources of correlations first are carefully looked for and then, if possible, eliminated before any large-scale association study is performed.

The following example illustrates the same general point, but now with 'cultural inheritance' as the source of the misleading correlation in a mixed population. Assume that a sample to be analysed comes from, say, a hospital with patients having different ethnic background. Assume also that the constituting groups are differentiated with respect to both an allele that is more common in one group than the other and a cultural difference leading to different risks for a particular disease. If in this mixed sample one looks for DNA markers correlated with the disease to hint at causative genetic variation, then the allele frequency difference between the groups may give a misleading signal of the existence of a genetic causative disease factor, when it only marks the correlation between cultural and genetic inheritance. It would, for example, be easy to find associations between native language and single nucleotide polymorphisms (SNPs) in a mixed group of Swedes and Chinese, in spite of the fact that no genetic variation exists that predisposes individuals for which language they will speak.

Different models of geographic structuration

Looking back at our main derivation, it is found to be based on a very specific and simplified model of how the geographic differentiation in a population looks. The assumptions originally outlined by Wahlund may be handy and analytically tractable, but real populations only rarely consist of completely separated subpopulations within which perfect random-mating occurs. This was, of course, known to Wahlund (1928), and in the latter part of his article there is an analysis of how migration between subpopulations with time will decrease the differentiation in the combined population.

Attention has, since Wahlund's time, been given to the question of what population model to use in attempts to determine the balance that evolves in large complex populations between genetic divergence caused by random drift between subpopulations and homogenization caused by migration between them. Let us here outline some different approaches, with increasing degrees of realism, which have been considered. Much valuable work on geographic differentiation was produced by Malécot (some of which has become more easily accessible by being reprinted in Malécot, 1966), but outside France it was the following papers that became influential and gave population genetics some of its key concepts.

Wright (1943) considered a population of diploids consisting of very many separate subpopulations of finite size exchanging a constant rate of migrants between one another in no structured pattern. In this 'island model', as he called it, he characterized the balance between migration and drift that will arise over time, and reached the general conclusion that the genetic divergence will remain limited as long as a subpopulation imports more than one migrant per generation. If the rate of migration is smaller than this, then genetic drift will make the subpopulations more substantially differentiated.

A more realistic way of modelling population differentiation is to give the subpopulations a true *geographic* structure, by letting subpopulations be in more migratory contact with 'neighbouring' subpopulations than with others. The simplest method to picture such subpopulations is to consider them as existing along a line (think: shore, river, road, mountain range) where they exchange migrants only with their immediate neighbours. This one-dimensional model can easily be extended to a two-dimensional structure where the subpopulations exchange migrants with the four subpopulations adjacent in the grid. Such 'stepping stone models', first analysed by Kimura and Weiss (1964), demonstrate the general importance of long-distance migration: a few migrants from subpopulations far away are much more effective in keeping the subpopulations relatively genetically homogeneous than an increased number of local migrants. Relative genetic homogeneity is also much more easily obtained in the two-dimensional than in the one-dimensional case. Yet one more step in this direction of realism was taken by Bodmer and Cavalli-Sforza (1968), when they described a matrix method to analyse genetic drift in a structure where there is a limited number of subpopulations with a specified migration rate between each pair. With this degree of detail, few general results can be obtained, and the value of the endeavour lies primarily in suggesting how prospective or retroactive numerical simulations studies may be performed, as outlined by Cavalli-Sforza and Bodmer (1971).

A fundamentally different way of modelling population differentiation was introduced by Wright, again in his article from 1943, when he considered what he called 'isolation by distance'. The idea is here that genetic drift occurs in a population where all individuals are evenly distributed over a continuous surface and migrate only a limited distance between birth and reproduction. The idea is elegant, but there are, of course, problems with assuming a continuous distribution of individuals that at the same time carry particulate genes and among which random drift occurs.

What we learn from this survey of population models is that it is never self-evident how to view the internal genetic structure of a population. When analysing data from real populations, a choice of which population model to use must be made, and it is in this choice that a scientist decides what balance between realism and structural simplicity to go for. And never can one hand this judicious choice over to a computer program or claim that one just uses the same assumptions and procedures as everyone else.

Additional measures to describe populations: F and F_{ST}

Earlier in this book, in Derivation 3, we described how the commonly used measures H and D can be interpreted in distinctly statistical ways. Here, we continue in the same mode by looking at F and F_{ST}, where we limit ourselves to the simplest possible situation of one locus with two alleles, A_1 and A_2, having frequencies p and q.

When there is some inbreeding in a population, we know from earlier that the frequencies of the three genotypes A_1A_1, A_1A_2 and A_2A_2 will be $p^2 + Fpq$, $2pq - 2Fpq$ and $q^2 + Fpq$. What does this situation say about the covariance in allelic state between the two gametes that unite in fertilization? Let X be the variable that takes value 1 if a male gamete is A_1 and 0 otherwise, and let Y be the corresponding variable for the female gamete. Obviously, the four possible outcomes for a combination (union) of these two variables – (1,1), (1,0), (0,1) and (0,0) – have the following probabilities: $p^2 + Fpq, pq - Fpq, pq - Fpq$ and $q^2 + Fpq$. The covariance in alleles between the uniting gametes thus becomes

$$
\begin{aligned}
Cov\,[X, Y] &= (p^2 + Fpq)(1 - p)(1 - p) + (pq - Fpq)(1 - p)(0 - p) \\
&\quad + (pq - Fpq)(0 - p)(1 - p) + (q^2 + Fpq)(0 - p)(0 - p) \\
&= p^2q^2 + Fpq^3 - p^2q^2 + Fp^2q^2 - p^2q^2 + Fp^2q^2 \\
&\quad + p^2q^2 + Fp^3q = \\
&= Fpq^3 + 2Fp^2q^2 + Fp^3q = Fpq(q^2 + 2pq + p^2) \\
&= Fpq = FV[X] = FH/2,
\end{aligned}
$$

where $V[X]$ is the variance of the variable X (equal of course to $V[Y]$) and H is the genetic diversity (the expected heterozygosity) at the locus. Since the standard deviation of X and of Y is the square root of the variance, we have that the correlation between the male and female gametes in allelic state is given by

$$
Corr\,[X, Y] = \frac{Cov\,[X, Y]}{SD[X]\,SD[Y]} = \frac{FV[X]}{SD[X]^2} = F.
$$

(For more on correlations, see Background 7:1.)

Thus, with no inbreeding in the population, there is no correlation between the gametes that meet in reproduction. We also see that the inbreeding coefficient of the population, F, earlier presented as the mean of the degree of inbreeding of its individuals, can alternatively be interpreted as the correlation in allelic state between the uniting gametes.

With respect to F_{ST}, we have already in expression (7.10) an explicitly statistical description of this measure with a clear interpretation: it stands for that part of the genetic variation (measured by H) that cannot be ascribed to the variation within subpopulations but instead depends on the variation (divergence) between them.

F_{ST} is best considered as measuring the distribution of a *single* allele over many subpopulations (just like D measures the relationship between a *particular* pair of alleles from different loci in the same population; see

Derivation 3). If there are many alleles at the locus, then they may each be associated with different F_{ST} values, just like different pairs of alleles may show different D values. How to make such estimates representative for the locus or loci involved has been much discussed. This topic belongs, however, to that part of population genetics that deals with producing good estimates from sampling data, which is very important but outside the scope of the present book.

Finally, let us note that we have here treated F_{ST} primarily as a measure of the genetic differentiation between subpopulations within a larger population. F_{ST} has also come to be used as a standard measure of what can be called the 'genetic distance' between two populations – their degree of distinctness as measured by some genetic variation. A hypothetical population is considered, made up of equal numbers from the two studied populations, and the F_{ST} value for this combined population is then calculated. The larger the F_{ST}, the more genetically different the two (sub)populations. Figure 7.1 shows how the F_{ST} value (and the total

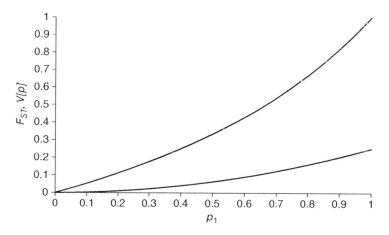

Figure 7.1 A population consists of two equally big subpopulations. Two alleles from the same locus are studied, A_1 and A_2. In the second subpopulation, allele A_2 is fixed. In the first subpopulation, the frequency of allele A_1, denoted p_1, varies from 0 to 1. The upper graph shows the increase in F_{ST} as the divergence between the two subpopulations increases as a function of p_1. The lower graph shows the genetic variance, $V[p]$, in the combined population as a function of p_1.

genetic variation) in such a situation increases as a function of the genetic divergence involved.

Why did he do it?

Wahlund's (1928) article on how to analyse population differentiation was published in *Hereditas*, a genetic journal edited from Lund in Sweden, where we – the two authors of this book – are active. Allow us therefore the right to briefly discuss the origin of Wahlund's analysis of population differentiation.

Sten Wahlund was born in 1901 and got his education from Uppsala University. In 1938, he became professor of statistics in Stockholm. He was politically active, represented the Agrarian party in the Swedish Parliament for 26 years from 1944, and died in 1976. He published only one paper in population genetics – the one we have used in the present chapter. It is clear and lucid, and the analysis he presents is new and path-breaking. No one had tried any similar analysis of population differentiation before his work in 1928, as far as we know.

When one reads his paper today, it is easy to get the impression that it is an exercise in formal genetics based on abstract ideas. That this is not fully correct is seen from the affiliation Wahlund gives: Staatsinstitute für Rassenbiologie, Uppsala (like most Swedes at this time, Wahlund published in German). The scientific problems underlying his work come, thus, from race biology and from the questions he studied at the Swedish Institute for Race Biology in Uppsala, under the leadership of the Swedish eugenicist Herman Lundborg. The stated task of the institute, at which Wahlund was chief statistician, was to improve (or at least prevent the deterioration of) the genetic stock of the Swedes; this was the explicit research direction during Lundborg's reign, which lasted from the start of the Institute in 1922 until his retirement in 1935.

The activity of Lundborg's institute is today much criticized in Sweden for its overt racism and links to Nazi Germany, and none of the work produced there is regarded as being of any particular value – except the important article by Wahlund.

Its links with the research interests of the institute are obvious. Most time and resources at the institute were devoted to physical-anthropological studies of the Sami people (who live in the north of Norway, Sweden, Finland and Russia; they were earlier often called the Lapps). Wahlund's interest in knowing if a sampled unit is to be regarded as homogeneous or not can in practice be translated into worrying about whether, for example, those speaking the different Sami languages can be taken to be genetically homogeneous or not (in Sweden today, three of these languages are used). Similarly, his interest in the gradual decline in genetic differentiation over time as a result of migration and intermarriages can be seen as a way to predict how long it will take before the Sami people disappears – something at least Lundborg regarded as inevitable.

There is no need to let the links with the Race Biology Institute or Lundborg and his contacts with Nazi Germany affect Wahlund's reputation, and his paper does not diminish in its scientific usefulness from these attachments. But the story reminds us – and the reader – that theoretical population genetics is not as abstract and unattached to the real world as it sometimes may appear. Behind the formula we discuss in this book can often be found different, and sometimes strongly diverging, views on race, eugenics and the effects of nuclear war. Or – as we will see in the next chapter – the roots to human sociability.

Background 7:1 More on variances, covariances and correlations

An introduction to means (expectations) and variances is given in Background 3:1. Here, we repeat and extend some of the results concerning variances. We also discuss the notion of *independence*, and introduce the concepts of *covariance* and *correlation*.

The variance of a distribution is always a positive number. If $V[X]$ is the variance of a variable X, and $Y = aX + b$ where a and b are constants, then

$$V[Y] = V(aX + b) = a^2 V[X]. \qquad (1)$$

This result implies, for example, that if we have a random variable p_i and define q_i as $1 - p_i$, then

$$V[q_i] = V[1 - p_i] = (-1)^2 V[p_i] = V[p_i].$$

Furthermore, the variance of a variable X is equal to the expectation of the square of the variable minus the square of the expectation; that is,

$$V[X] = E[X^2] - E[X]^2, \qquad (2a)$$

which in the notation already used in this chapter can also be written as

$$V[X] = \overline{x^2} - \overline{x}^2. \qquad (2b)$$

In Background 3:1, we discussed how the expectation of the sum of two random variables equals the sum of their expectations:

$$E[X + Y] = E[X] + E[Y].$$

We cannot expect the same simple relationship to hold with respect to the variance. If, in some way, the two variables are associated so that when X takes a low value, so does Y, while when X takes a high value, so does Y, then the variance of the sum of the variables must be greater than the sum of their respective variances. This is where the notion of the *covariance* between two random variables enters.

The covariance is a measure of how two random distributions covary with each other. It is based on the distance between the values of the distributions and their means in the following way:

$$Cov\ [X, Y] = E[(X - E[X]) \cdot (Y - E[Y])].\qquad(3)$$

With a rewriting, similar to what has just been done with the variance, it can be shown that the covariance may be expressed like this:

$$Cov\ [X, Y] = E[XY] - E[X]E[Y].\qquad(4)$$

When the two distributions are – to use the statistician's term – *independent*, then $E[XY] = E[X]E[Y]$, and, thus, the covariance between them is zero.

We also accept without proof that

$$V[X + Y] = V[X] + V[Y] + 2Cov\ [X, Y],\qquad(5)$$

which can be used to tell us that if the variables are independent, then

$$V[X + Y] = V[X] + V[Y].$$

Finally, the *correlation coefficient* between two distributions, say X and Y, is defined as:

$$Corr\ [X, Y] = Cov\ [X, Y]/\sqrt{V[X]\ V[Y]} = \frac{Cov\ [X, Y]}{SD[X]\ SD[Y]}.\qquad(6)$$

It is thus the covariance between the distributions standardized by their respective spreads, measured by the standard deviations.

A final note on word use: When, say, an allele occurs together with a trait value more often than could be expected by chance, we prefer to call them 'positively associated'. The term 'correlated' we restrict for more technical usage. And never do we say that the allele and the trait value are 'linked', since this may lead to confusions given the very specific meaning 'linkage' has in genetics.

Derivation 7

Questions

1. Assume a population consisting of two subpopulations, where the first is three times as large as the second. Two alleles, A_1 and A_2, segregate in the population; the frequency of A_1 is 0.7 in the first subpopulation and 0.5 in the second. Calculate $\bar{p}$, $V[p]$, H_S and H_T for this situation. Check numerically that $\frac{V[p]}{\bar{p}(1-\bar{p})}$ equals $\frac{H_T-H_S}{H_T}$, in accordance with the two ways of calculating F_{ST} as described by expressions (7.6) and (7.10).

2. Assume a haploid organism. Further assume that two loci, A and B, segregate in two subpopulations so that $p(A_1) = 0.7$ and $p(B_1) = 0.4$ in population I, whereas $p(A_1) = 0.5$ and $p(B_1) = 0.1$ in population II. The two populations are of equal size. Within each subpopulation, there is no linkage disequilibrium. What is the expected level of linkage disequilibrium, D, if the two populations are combined in equal proportions?

3. Assume that in a geographic area, some local human communities have been separated from one another for such a long time that they have become slightly genetically differentiated. In less than one generation, this separation disappears due to better communications, and the inhabitants can nowadays be regarded as random-mating. How rapidly is the Wahlund effect lost, so that the joint population shows Hardy–Weinberg proportions? Will any linkage disequilibrium due to the mixing of the subpopulations be lost at the same rate?

4. The three alleles A_1, A_2 and A_3 have frequencies 0.4, 0.4 and 0.2 and 0.1, 0.1 and 0.8 in two equally big, random-mating subpopulations. What are the genotype frequencies in the combined population and what frequencies are expected under the assumption that the combined population had been produced by random mating? Compare these values: Any surprises?

5. Assume that from a large haploid population, where the frequency of allele A_1 is p, a number of new populations are formed by repeatedly sampling n haploids at random from the original population. What is the expected F_{ST} among these newly formed populations?

Derivation 8

Cost and benefit of sib-interactions

The genetics employed in population genetic analyses is normally straightforward: populations are well-defined and genes are reasonably stable and heterozygotes segregate their alleles as envisaged by Mendel. This simplicity gives the field its unity and intellectual rigour, but the solid foundation should not fool us. Perhaps the most important factor in population genetics – selection – is almost never simple. Selective forces may act in many different ways and through the most intricate paths.

The three final derivations deal with the effect of selection, allowing for more complex actions and interactions than assumed in the earlier chapters. We start by studying the case where a well-defined selective logic exists, but where it is not easy to say whether it acts in favour a particular allele or not. The allele we consider decreases the immediate fitness of its carriers but improves the fitness of the broods in which it occurs – so what will happen in evolution? Will the mutation spread when new and therefore rare?

A useful approach to the study of such complicated situations is to consider a model in which the effect under analysis is made highly simplified and explicit. The model gives perhaps only a caricature of the real world, but within its framework population genetics may determine the expected long-term evolution of the considered effect, and thereby elucidate what is possible in evolution and what is not. Such an analysis becomes similar to the 'thought experiments' commonly used in physics, where realism and details are left aside to allow for an understanding of what consequences the logic in a set of assumptions ultimately leads to.

Understanding Population Genetics, First Edition. Torbjörn Säll and Bengt O. Bengtsson.
© 2017 John Wiley & Sons Ltd. Published 2017 by John Wiley & Sons Ltd.

In the latter parts of the chapter, we discuss how more general forms of group-related selection (and not just sib selection) may favour the evolution of altruistic traits, crudely defined as traits detrimental in fitness to those that exhibit them but advantageous to those that receive the benefits from them. Then we consider other types of selection where a genotype's fitness depends on the genetic composition of the surrounding population, generally called *frequency-dependent selection*. And we finish with a discussion of the selective forces that act on genes involved in resistance reactions towards pathogens.

Analysis

The question that we use as a starting point for our analysis of the complex possible effects of selection, is whether an unselfish trait may evolve in evolution despite the fact that natural selection acts on individuals and, therefore, would appear to be restricted to advance traits favourable to individuals. In our analysed model, we test whether the unselfish behaviour of an individual that brings fitness benefit to its sibs but not to the rest of the population may provide the possibility for an altruistic trait to evolve.

Assumptions

In a very large, random-mating population of a diploid organism with males and females in equal number, there is a locus A with standard allele A_1 and a new rare dominant mutation A_2 that affects the behaviour of the young while they still interact with their siblings. Heterozygous carriers of this allele are altruistic to their sibs (irrespective of genotype) at a certain risk to their own life. This implies that they will be at a relative disadvantage in sib-ships with A_1A_1 brothers and sisters, but that sib-ships with at least some A_2 carriers will on average be larger than those without any A_2 carriers. A specific behaviour (danger-warning) that may lead to this kind of situation is examined later.

We start, however, by considering the general situation and assume that in a sib-ship born to a couple with one parent having the new altruistic allele, the brood ends up with A_1A_1 and A_1A_2 offspring in a ratio of $1 : (1 - s)$, instead of the expected ratio $1 : 1$. The final size of this sib-ship is, on the other hand, $1 + t$ relative to the standard size of sib-ships lacking the new allele, which we take to be 1. The allelic variation has no other selective effects and the genotypes at birth are in Hardy–Weinberg proportions. Thus, for s and t strictly positive, this describes a situation where the new allele A_2 can be taken to have both good and bad fitness effects. These assumptions (similar to the ones made by Williams and Williams, 1957) lead to an analysis that is mathematically very simple – next to trivial. We pay instead particular attention to the recurrence system, which differs from the standard procedure used in the earlier derivations, but which opens up for a fruitful general method by which to study complex evolutionary questions (further discussed in Derivation 10).

Finding a relevant recursion equation

Of great help to our analysis is the assumption that we only look at the considered mutation when it is newly introduced; that is, when it is very rare. As we will see, this makes it possible to drastically reduce the complexity of the calculations.

Since A_2 is rare, almost all copies of the new mutation will be carried by heterozygotes (see Background 2:1); let the frequency at the time of mating of such heterozygotes be y. Homozygotes for the allele will be so rare that we can ignore them. This leaves us at the time of mating with only the heterozygotes plus the standard homozygotes with their frequency $1 - y$, which we denote x. Since all terms of size y^2 can be ignored (they contain squares of the rare allele frequency), we only need to consider two types of mating: those between two normal homozygotes, and those between one normal homozygote and one heterozygote. This gives us the following much reduced mating type/offspring table:

			Offspring		
Mating type	Frequency	Sib-ship size	A_1A_1	A_1A_2	A_2A_2
$A_1A_1 \times A_1A_1$	x^2	1	1	0	0
$A_1A_1 \times A_1A_2$	$2xy$	$1+t$	$1/2$	$1/2(1-s)$	0

That the table is complete, given our assumptions, is seen by adding the frequencies of the mating types. Their sum is $x^2 + 2xy$, to which we can add the very small term y^2, making the total into $(x + y)^2 = 1$. We see that the assumptions about selection acting against A_2 within mixed sib-ships, s, as well as the selective advantage of such mixed sib-ships relative to the standard sib-ships, t, are both represented in the table.

This mating table could be used to calculate the expected frequency of allele A_2 in the new generation, but we continue to do the analysis based on genotype frequencies and not allele frequencies. The frequencies of the two relevant genotypes after selection (*i.e.* after the broods have dispersed) are given by

$$x' = [1 \cdot 1 \cdot x^2 + (1 + t) \cdot \tfrac{1}{2} \cdot 2xy]/W = [x^2 + (1 + t)xy]/W$$

$$y' = (1 + t) \cdot \tfrac{1}{2}(1 - s) \cdot 2xy/W = [(1 + t)(1 - s)xy]/W, \qquad (8.1)$$

where

$$W = x^2 + (1 + t)xy + (1 + t)(1 - s)xy$$

$$= x^2 + xy + txy + xy - sxy + txy - stxy$$

$$= x^2 + 2xy + 2txy - sxy - stxy = (x + y)^2 - y^2 + 2txy - sxy - stxy$$

$$= 1 - y^2 + (2t - s - st)(1 - y)y \approx 1 + (2t - s - st)y.$$

Here, we have used the same method as in Derivation 2, where the effect of fixed selection values is studied. To formulate a recursion system that describes what happens between two consecutive generations involving selection, it is necessary to get the selected proportions expressed as relative frequencies (so that they add to 1), which is obtained by dividing with the corresponding normalizing sum. (In the present case, the value W cannot, however, as easily be given a meaningful interpretation as in Derivation 2, where the W value equals the mean fitness of the population.) This technique, based on a general normalizing factor W, allows also for a greater flexibility in the construction of mating-offspring tables.

As in the analysed case of heterozygote advantage, we only need to follow the evolution of y, since when we know y, we always know x. The relevant recursion describing the dynamics of our system is, thus,

$$y' = (1 + t)(1 - s)xy/W = (1 + t)(1 - s)(1 - y)y/W$$
$$\approx (1 + t)(1 - s)y/[1 + (2t - s - st)y],$$

from which we, with a natural approximation (see Background 2:2), obtain

$$y' \approx (1 + t)(1 - s)y. \tag{8.2}$$

The condition for an increase of the new, altruistic allele A_2 has thereby been found. Whether the allele will spread or not depends on whether $(1 + t)(1 - s)$ is greater or smaller than 1 – which can be expressed as saying that the advantage to the sib-ship in which some offspring carry this allele must be greater than the disadvantage to the carriers of the allele within these sib-ships. This is a correct way to express our results, but it should be remembered that recursion (8.2) describes the frequency of heterozygotes, y, and not the frequency of allele A_2 itself. This is per-fectly all right, however; there is no rule in population genetics which says that all evolutionary dynamics must be expressed in allele frequencies alone.

Danger warning

We will now give a somewhat more concrete illustration of what the obtained result means. The example is still contrived, but it exem-plifies an altruistic behaviour of the kind that we want the model to capture.

Let the size of all sib-ships at birth be n (which is not a small number). Assume that every sib-ship is attacked exactly once by a predator. In the first mating case ($A_1 A_1 \times A_1 A_1$), exactly a offspring die from this attack. In the second mating case ($A_1 A_1 \times A_1 A_2$), one offspring with genotype $A_1 A_2$ warns its sib-mates (we assume that all such sib-ships are so big that they always contains at least one with this genotype) and saves them all – except that it itself is killed in the process. How big must a be to make this altruistic act favoured in evolution?

The number of offspring in the different cases can be summarized as follows:

Mating type	Frequency	Sib number	$A_1 A_1$	$A_1 A_2$	$A_2 A_2$
		Offspring			
$A_1 A_1 \times A_1 A_1$	x^2	$n-a$	$n-a$	0	0
$A_1 A_1 \times A_1 A_2$	$2xy$	$n-1$	$n/2$	$(n/2)-1$	0

We can now link this table, where the offspring are described in numbers, to the preceding table, where the description instead is in relative proportions. This comparison gives us

$$1 - s = [(n/2) - 1]/[n/2] = (n - 2)/n$$

and

$$1 + t = (n - 1)/(n - a).$$

By the use of expression (8.2), the condition for spread of the altruistic behaviour in the present case can then be written:

$$(1 + t)(1 - s) > 1 \iff \frac{(n - 1)}{(n - a)} \frac{(n - 2)}{n} > 1 \iff$$

$$\frac{n^2 - 3n + 2}{n(n - a)} > 1 \iff \frac{n - 3 + \frac{2}{n}}{n - a} > 1 \iff$$

$$n - 3 + \frac{2}{n} > n - a \iff a > 3 - \frac{2}{n}. \tag{8.3}$$

Thus, a sufficient condition for the spread of the altruistic behaviour is that a is equal to or greater than three (n is assumed to be large). With this number lost from standard-sized sib-ships, the presence of the altruist allele implies that two offspring (*i.e.* 3 − 1) are saved in every sib-ship belonging to the second type of mating. A classical way to express this result is to say that evolution favours that one sib dies from its altruistic behaviour if it thereby saves two or more of its sib-mates (an adage supposedly due to Haldane, and expressed by him, for example, in 1955).

Summing up

For some traits, the fitness of an organism depends on the context in which it lives and what other genotypes it interacts with. It may then not always be immediately clear if a particular allele which influences this trait is favoured by selection or not. Population genetics provides tools and a practical framework for such analyses, and in this derivation we show how sib selection can be investigated. In a highly simplified example, we illustrate how an allele with a trait that makes its carriers risk their life will be selectively favoured if only the carriers thereby on average save at least two of their full sibs.

With this, we have also seen how highly simplified models in population genetics – in situations where a detailed fit to reality is far too complex to achieve – still may produce important insights via a kind of 'proof of principle'.

Interpretations, extensions and comments

Altruism and natural selection via individuals and groups

Charles Darwin (1859) regarded the selective forces that the environment imposes upon an organism as the most important agents for its evolutionary change, be these factors physical or biological. While arguing for this idea, he did not discriminate between forces that primarily act on individuals directly and those that act on groups of individuals. However, when his theory of evolution became recast within the genetic framework, this question of individual versus group selection grew more acute. Can, for example, natural selection acting on individuals explain the evolution of what may be called 'altruistic traits', when altruism – however vaguely conceptualized – always implies a behaviour unfavourable to the individual though favourable to others? Or is altruism a trait that can only be explained via selection at the collective level, the group level?

Even though group selection was quietly accepted as an important factor by most evolutionary thinkers for more than one hundred years, both Fisher and Haldane found it relevant to point out that a particular version of individual selection via siblings may, at least in some instances, explain the evolution of traits negative to their carriers. Fisher (1930) discussed the origin of the prominent colours that many distasteful insect larvae show, and suggested that the death of one larva may lead to the survival of its nearby sibs, since the attacker will associate the disagreeable taste of the eaten larva with the distinct colour. Haldane (1955) extended this argument when he discussed the evolution of acting heroically (saving a relative from drowning at a personal risk) and explained that this logic will function not only with full sibs but also with other more distant relatives, though then with less efficacy.

The English evolutionary biologist William (Bill) Hamilton (1963) gave kin selection as an explanation of altruism its explicit foundation with his well-known formula

$$k > \frac{1}{r},$$

saying that altruism will spread if the gain in fitness due to altruism of a relative compared to the donor (k) is greater than the inverse of their degree of relatedness (r). The strength of this line of thought became obvious when he in 1972 explicated how his rule applies to animals with a haplodiploid sex determination system. Among these, to which belong the social insects of the order Hymenoptera, such as ants, wasps and bees, the sterile workers are unusually highly related to the sibs they raise, which helps explain these workers' unusual behaviour.

With the insight that traits like altruism, as well as other types of social behaviour, can be approached with strict evolutionary methods, optimism spread that all kinds of social behaviour were ripe for Darwinian analysis. The resulting debate on the relevance of a human socio-biology, starting in the 1970s, did not introduce many ideas of direct interest for population geneticists, but it did generate at least a valuable interest in such subtle evolutionary processes as kin selection.

It should, however, be realized that the key to our derivation using sib selection, and to the explanations given by Fisher, Haldane and Hamilton, is not to be found in relatedness *as such*. What relatedness provides is a positive association between the alleles carried by the donor and the receiver of the altruistic act. Altruistic traits may evolve wherever such positive associations exists, if only the balance between the factors involved is suitably weighed. Thus, an allele for altruism (or any similar trait) will spread if and only if sufficiently many beneficiaries of the behaviour carry the allele in question. Seen in this way, sib selection becomes just an intermediary point in a continuum of selective processes ranging in their actions over individuals to groups (as described by Uyenoyama and Feldman, 1980).

The *sine qua non* for all group-selective processes to function is, however, that there are distinct and long-lasting allele differences between the groups. As usual in evolution, if there is no heritable variation, then selection will carry no evolutionary effect. Any migration between groups that leads to an equalizing of the allele frequencies between them will therefore always make group selection difficult. This was illustrated by Maynard Smith (1964) in a classic biological thought-experiment assuming mice colonizing haystacks. As long as migration between the haystacks is small, 'timid' mice that produce many offspring at the end of the reproductive season will outcompete 'aggressive' mice that have an individual selective advantage but produce a smaller reproductive outcome. With a small amount of migration between the haystacks, the timid mice win, but with increasing migration, the aggressive mice become increasingly favoured by evolution.

Thus, the negative effect that some behaviour has on the fitness of those that perform it may be compensated for by the positive fitness effect that others gain from such behaviour. The situations in which this selection scenario is relevant are, however, highly restricted, particularly in organisms practising regular outbreeding. From this perspective, the ultimate example of group selection therefore becomes the demise or success of *species* over time, since these exemplify evolutionary units that – by definition – are reproductively isolated and between which allele frequency differences certainly exist.

Frequency-dependent selection and resource competition

Let us now leave the problematic issue of how social traits evolve and reconnect with the more general question raised in the introduction to this chapter regarding the complex nature of selection. In our earlier discussions of selection in Derivations 2 and 6, it was assumed that the fitness values associated with the different genotypes are fixed. This basic assumption is sometimes perfectly reasonable: a lethal dominant allele will always give the heterozygotes a fitness of 0, irrespective of circumstances. But often it is more natural to consider a genotype's fitness as a dynamic function, changing with – among other factors – the frequencies of all the genotypes in the population.

We illustrate the dynamic nature of genotypic fitness values by considering what happens if selection due to resource competition is made to function within a population genetic framework. Our description is limited to the derivation of a relevant recursion system, since we just want to illustrate what such a system may look like.

In ecology, there is a tradition of describing the population dynamics of a species via a *logistic equation*. When partially competing species are present, the demographic evolution of a species x may be described by an equation of type

$$N_x' = N_x \left[1 + R_x \left(1 - \frac{N_x + \alpha_{x,y} N_y + \alpha_{x,z} N_z + \ldots}{K_x} \right) \right].$$

In this recursion, N_x is the number of individuals in species x in the current generation, R_x is its maximal reproductive rate and K_x is its carrying capacity. The subscripts y and z denote other, partially competing species, and the α values measure the resource competition between them (which, in our treatment here, is assumed to be symmetric, *i.e.* that the effect of the competition between an individual of x with respect to an individual of y equals the effect in the opposite direction). The prime sign, $'$, denotes the new generation.

This way of modelling selection can readily be adapted to situations where there is competition between genetically different individuals *within* a species. (Here, we simplify but follow the arguments of Christiansen and Loeschcke (1980), based on Christiansen and Fenchel (1977).) Assume, for example, that in a predator species there is a locus A with two alleles, A_1 and A_2 that affect what prey their carriers eat. The predators with genotype A_1A_1 eat small prey, those with genotype A_1A_2 eat intermediary-sized prey and those with genotype A_2A_2 eat large prey. The frequency of A_1 is p, while the frequency of A_2 is q (where $p + q = 1$).

If there is random mating in the population, then we can use the ecological model just described to formulate a recursion for the demographic dynamics of the predator species in the following way:

$$N_{11}' = p^2 N \left[1 + R_{11} \left(1 - \frac{N_{11} + \alpha_{11,12} N_{12} + \alpha_{11,22} N_{22}}{K_{11}} \right) \right]$$

$$N_{12}' = 2pq N \left[1 + R_{12} \left(1 - \frac{\alpha_{11,12} N_{11} + N_{12} + \alpha_{12,22} N_{22}}{K_{12}} \right) \right]$$

$$N_{22}' = q^2 N \left[1 + R_{22} \left(1 - \frac{\alpha_{11,22} N_{11} + \alpha_{12,22} N_{12} + N_{22}}{K_{22}} \right) \right]$$

$$N' = N_{11}' + N_{12}' + N_{22}',$$

Here, N is the total number of individuals in the population and N_{11}, N_{12} and N_{22} are the numbers of individuals with genotypes $A_1 A_1$, $A_1 A_2$ and $A_2 A_2$; obviously, $p = (N_{11} + \frac{1}{2} N_{12})/N$ and $q = (\frac{1}{2} N_{12} + N_{22})/N$. The other parameters have the same meaning as before, but the subscripts now specify genotypes rather than species. The α values describe the resource competition between the different genotypes, and it is therefore natural to assume that all α values fall between 0 and 1, which implies that the genotypes certainly compete for partially common resources, but that the strongest competition is always between organisms with the same genotype. Our equations, thus, describe a model of *reduced* competition, since individuals with different genotypes compete less with each other than they do with genotypically identical individuals.

In this recursion system, it is clear that the expressions within the square brackets function as the fitness values for the different genotypes. And these fitness values turn out to be quite complex: not only do they depend on a number of parameters like the R, K and α values, but they depend also on the numbers and frequencies of the other genotypes in the species. We are here very far from the assumptions about fitness values being fixed, as normally made in elementary population genetics.

At the same time, the equations do not say anything strange. If, in a predator population where allele A_1 is standard, a new allele A_2 enters, giving its carriers a larger mouth, bill or general feeding apparatus and thereby an opportunity to eat larger prey, then this allele will spread if and only if there is an unused abundance of such prey available. The equations are, thus, highly suited for describing the dynamics taken by a population when it genetically adapts to its ecological niche.

They may also lead to results that seem perfectly natural in this more complex setting, but which would be unthinkable given fixed fitness values. If there are prey of all sizes, then the predator population will certainly show a genetic polymorphism at the A locus, since the A_1 allele as well as the A_2 allele will be favoured when rare. But at the internal equilibrium, towards which the population will move, the heterozygotes A_1A_2 may well have the lowest fitness, since they must compete with both types of homozygotes for the prey that they are best adapted to. Thus, we see here that in a population there may very well exist a genetically balanced polymorphism at which heterozygotes are at a relative *disadvantage* – something which is never seen when fitness values are fixed, as proven in Derivation 2. An illustration of this possibility is shown in Figure 8.1.

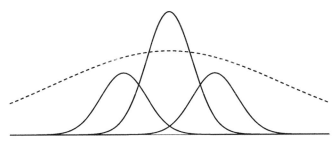

Figure 8.1 Competition between genotypes in a predator species for a continuously distributed prey (dashed graph). A symmetric situation is assumed, where the three genotypes (A_1A_1 to the left, A_1A_2 in the middle and A_2A_2 to the right) differ in their resource use. Genetic variability is obviously favoured, since the alternative allele will always increase in frequency if introduced when the other allele is fixed. Equally obvious is that the heterozygotes are at a selective disadvantage in the illustrated equilibrium situation, since competition is most severe for their preferred resource. (For formal analyses of situations like this, see Christiansen and Loeschcke, 1980.)

Rare allele advantage due to infections and self-incompatibility

For another example of new and spectacular genetic effects that may follow from frequency-dependent selection, let us consider the evolution of genetic responses to infections.

If one studies the distribution of genetic variation over the human chromosomes, the most striking feature is not any troughs in the level of variation indicating putative selective sweeps (as discussed in Derivation 6), but an incredibly high peak in variation over a restricted part of the short arm of chromosome 6. Here, the human leukocyte antigen (HLA) region is situated – the human version of what in animals generally is called the major histocompatibility complex (MHC). To describe the enormous variation at this site of about 1 Mb length, corresponding to about 1 cM, we can tell that at the time of writing the A, B, C and DRB1 genes in the region code for 2 313, 3 011, 1 985 and 1 335 known different proteins, respectively (Robinson *et al.*, 2015).

What is the cause of this sensational variation? Why does the balance between mutation and drift, discussed in Derivation 5, lead to such a very different result here? The answer is, of course, provided by the special kind of selection that acts on new allelic variants of HLA.

All well-functioning infectious agents, be they viruses, bacteria or fungi, are good at rapidly adapting themselves to the host they colonize. The best response to attacks by such agents is therefore to be *different*. In this case, this means to be different at the cellular level from most other members of the species, so that an infectious agent that, for example, uses a common molecular structure for its attack on cells will lack this possibility. New variants of genes coding for cell-surface proteins (such as the four genes cited in the previous paragraph) will thus be associated with selective advantages, since their carriers will be infected less often than individuals with common cell-surface types. This advantage for new alleles will, however, only persist as long as the alleles are rare, because when these alleles become common, the infectious agent will rapidly adapt to their proteins.

Thus, we find here yet another natural example of frequency-dependent selection, in this case arising from rare alleles having a fitness advantage relative to more common ones. Such situations will, of course, lead to high levels of variation for the genes involved.

The genetics of HLA is highly complex, and there can be other causes for its richness of genetic variation than the one just outlined. That the logic of rare allele advantage may lead to high levels of local genetic variation is, however, illustrated by another example arising in a completely different biological situation:

Many angiosperm plants have a genetic self-incompatibility system, known as a gametophytic system, which prevents pollen from the plant from fertilizing the plant itself. It does so through a system whereby a pollen grain carrying a particular self-incompatibility allele cannot grow on the stigmas of plants heterozygous for this very allele (with this system, no homozygotes for the specificity can ever arise). Such self-incompatibility loci are always highly variable, with even small populations containing a large number of alleles. The reason is obviously the same as discussed for HLA. Rare alleles can, via pollen, fertilize almost every other plant in the population, while common alleles will be at a reproductive disadvantage in that they will not be able to fertilize as many plants.

Thus, selecting for rare alleles gives the same result – a local inflation in genetic variation – in two very different biological situations. This shows that even when selection acts in intricate ways and due to completely different causal mechanisms, it may lead to evolutionary results of highly similar structure.

Questions

1. Return to the analysed danger-warning situation and give verbal answers without any calculations: (a) How is the outcome affected if some sib-ships are produced by females mated by more than one male? (b) Does it matter if there are some monozygotic twins among the siblings? What about dizygotic twins?

2. In a random-mating population, there are two alleles at locus A, A_1 and A_2, with allele frequencies p and q. Homozygotes A_1A_1 have constant fitness 1, while A_2 is strictly dominant and gives to A_1A_2 and A_2A_2 the frequency-dependent fitness of $1 + a - bq$, where $0 < a < b < 1$. Describe the evolutionary dynamics at this locus without doing any explicit calculations (and then *with* calculations, if you so wish).

3. According to the data given for the variability of the four loci in the HLA region, how many diploid genotypes (in terms of expressed proteins) can these genes produce?

4. Consider a population with a self-incompatibility locus of the type discussed at the end of this chapter, but with only three alleles, S_1, S_2 and S_3. In such a gametophytic self-incompatibility system, there can only be three genotypes: $S_1 S_2$, $S_1 S_3$ and $S_2 S_3$. (Convince yourself of this!) Let a, b and c denote the relative frequencies of these genotypes in the population. Pollen production and seed-set are equal for all individuals, and the different genotypes have identical survival rates. A very large amount of pollen is produced, and the fertilizations allowed by the incompatibility system are random.

   Derive the recursion equations for the genotype frequencies. Then let $a_0 = 0.9$, $b_0 = 0.1$ and $c_0 = 0$ and calculate the genotype frequencies in the three following generations. Guess what the equilibrium frequencies for the alleles will be, and show that your guess in fact corresponds to an equilibrium point.

Derivation 9

Selection on a quantitative trait

Selection acts on traits, not directly on alleles or genotypes. And the majority of traits under selection are quantitative in nature, concerning properties like length, size and weight, but also behaviour and physiological states. One question this raises for population genetics – with its primary interest in allele frequencies – is what effect selection on a quantitative phenotype has on the underlying genetic variation. Strong phenotypic selection does not automatically translate into strong selection for and against alleles, if the genetic influence on the trait is weak. But what happens when there is a considerable genetic component to the trait and many alleles at different loci affect the trait value – how strong, then, will selection be at the participating loci?

When selection acts on a quantitative trait, two effects are produced. The first is that the frequencies of the underlying genotypes shift, and with them the allele frequencies in the population. How allele frequencies change due to selection on genotypes has been well analysed and studied since the start of population genetics. All textbooks present models for different selection schemes and give results for what they lead to; we have done so in Derivation 2.

The second effect is that there is a shift in phenotypes in the selected population. If selection is unidirectional, then this shift will be seen as a change in the trait mean (which later may be continued into the next generation). Such phenotypic effects are traditionally presented and discussed in *quantitative genetics*, a branch of genetics closely connected

Understanding Population Genetics, First Edition. Torbjörn Säll and Bengt O. Bengtsson.
© 2017 John Wiley & Sons Ltd. Published 2017 by John Wiley & Sons Ltd.

to animal and plant breeding, and therefore particularly interested in phenotypic changes.

How shifts in trait means relate to the underlying allele frequencies is, however, seldom studied and rarely made explicit. Actually, one may say that the whole point of quantitative genetics is to be able to analyse heritable phenotypic changes *without* having to consider what happens at the genetic level in explicit detail.

This is, however, what we now shall do. We ask how strong selection will be at a locus with two alleles, as a function of the selection that acts on a quantitative trait affected by the locus. We start with a brief introduction to quantitative genetics modelling (following the treatment in Falconer and Mackay, 1996), before we recapitulate our earlier derived results about selection on genotypes. In the derivation, we stay close to classic quantitative genetics, though at one point we use a nice but not very common formula concerning the mean of truncated normal distributions, which is formally proved in Background 9:1.

The result of our derivation leads to a discussion of what selective dynamics to expect for alleles influencing quantitative traits. When is the evolutionary behaviour of such alleles ruled by the selection applied, and when do the alleles evolve primarily due to genetic drift?

A disease – or any other 'either/or state' – cannot normally be regarded as a quantitative trait. We therefore present the classical method based on liabilities, which makes such traits amenable to quantitative genetic analyses. The concluding evaluation of the assumptions used in quantitative genetics leads to the chapter's open end: Recent genomic studies show that these assumptions are generally valid and sound. This implies – paradoxically – that the detailed genetic causation of many important traits will be hard to establish.

Analysis

Selection in quantitative genetics

To explain how quantitative genetic analysis functions, we assume a trait where the phenotype can be measured on a quantitative scale. It is further assumed that the phenotype is influenced by allelic variation at a number of loci, but also by the environment, and that these influences affect individuals in an independent manner. The population is assumed to be diploid, random-mating, very large and homogeneous, in that every individual may have any genotype and encounter any environment. The phenotype of an individual i, P_i, can then be described by the simple relationship $P_i = G_i + E_i$. Here, G_i is the effect of the genotype of the individual and E_i is the environmental influence that the individual has experienced.

In the following, we will be particularly interested in the variance of the phenotypes in the population, σ_P^2, which under the assumption of independence between G_i and E_i becomes

$$\sigma_P^2 = \sigma_G^2 + \sigma_E^2.$$

This follows from the standard statistical result that the variance of a sum of two independent variables is the sum of the variances of the two variables (see Background 7:1); if the variables are not independent then this expression will include a covariance term and everything becomes more complicated. In consequence of this notation, we use σ_G^2 for the variance due to genetic variation and σ_E^2 for the variance in the population caused by individuals being differently affected by the environment, independent of their genotypes.

Let the mean phenotype in the population be μ_{Orig}. The subscript shows that this value is the trait mean in the *original* population, and we will now follow what happens when directional selection is applied to this trait. We assume that only a fraction – a selected sample – is retained in the population after selection (to later be allowed to breed and produce the next generation). The phenotypic mean of the selected part of the population we denote μ_{Sel}. The standard way in quantitative genetics to express the *strength of selection*, usually denoted by S, is to let it be the difference between these means:

$$S = \mu_{Sel} - \mu_{Orig}. \tag{9.1}$$

The strength of selection, S, will, thus, be measured in the same unit as the trait itself, say cm or kg.

For our present purpose of linking this phenotypic selection to the selection at the genetic level, we do not need to follow what effect the assumed selection will have on the trait in the next generation, but let us out of completeness outline what to expect.

If the individuals in the selected fraction of the population reproduce after random mating and produce a new generation with mean phenotype μ_{Next}, then the *response* to selection, R, is in a natural way measured by:

$$R = \mu_{Next} - \mu_{Orig}.$$

To relate this response, R, to the strength of selection and the trait variance in the original population, the following partitioning of the genetic variance is useful:

$$\sigma_G{}^2 = \sigma_A{}^2 + \sigma_D{}^2.$$

Here, $\sigma_A{}^2$ is the so-called *additive genetic variance*, which measures the amount of genetic variation due to additive effects in the original selected population. The rest of the genetic variance is normally called the *dominance variance* and is denoted $\sigma_D{}^2$. This is the part of the genetic variation that cannot be used by immediate selection to change the population mean. (Central to quantitative genetic analysis is to figure out how to define, and then to measure, these variance components, which is often both cumbersome and difficult.)

The additive genetic variance, $\sigma_A{}^2$, is of fundamental importance for the effect that selection has on the trait, and a key result in quantitative genetics theory is that

$$R = \frac{\sigma_A{}^2}{\sigma_P{}^2} S. \tag{9.2a}$$

This result is presented and discussed in most elementary text books. It states that the response to selection is a direct function of the strength of selection and the proportion that the additive variance constitutes of the total phenotypic variance. The proportion $\frac{\sigma_A{}^2}{\sigma_P{}^2}$ is generally called the *heritability* (in the narrow sense) of the trait, and it is classically symbolized by h^2. Another way to express the standard rule of quantitative genetics is, thus, to write

$$R = h^2 S, \tag{9.2b}$$

which is commonly called the breeder's formula.

We are, however, not here interested in what phenotypic effects one or more rounds of selection will have on a population over generations, but in the effect one such round of phenotypic selection has on the underlying genetic variation that contributes to the variation in the trait. We therefore now return to our main argument.

Selection in population genetics – one more time

Here, we repeat what has already been discussed in Background 2:2, since it is so central to our present concern.

Assume a locus, Λ, with two alleles A_1 and A_2, where the three genotypes A_1A_1, A_1A_2 and A_2A_2 have probabilities w_{11}, w_{12} and w_{22} to survive to reproductive maturity (*i.e.* to be retained after the act of selection). We assume that A_1 is the favoured allele and that it is neither strictly dominant nor recessive, which implies that $w_{11} > w_{12} > w_{22}$. No other differences with respect to fitness are assumed to exist between the genotypes.

When the effect of the differential survival on the allele frequencies is calculated, only the relative sizes of the survival probabilities matters. Because of this, the survival probabilities can be scaled so that the relative fitness of A_1A_1 is $w_{11}/w_{11} = 1$, that of A_1A_2 is $w_{12}/w_{11} = 1 - hs$, and that of A_2A_2 is $w_{22}/w_{11} = 1 - s$. Expressed like this, s is called the *selection coefficient* for the locus and h the *degree of dominance* of the A_2 allele (this h is different from the parameter h used in expression (9.2b); no confusion will arise, however, since as long as the h introduced here falls between 0 and 1 – as, according to our assumptions, it should – it does not explicitly enter any of the future calculations).

Notations and assumptions

We have now introduced two parameters that quantify the strength of selection in the outlined situation, S and s. It is inconvenient that the same letter is used to denote both parameters, but textbooks consistently use this notation and therefore we do not introduce any other symbols. The aim of our derivation is to find the relationship between these two parameters.

To reach our aim, we need to make a number of assumptions. Let us first consider those that involve the phenotypic variation and selection of the population as a whole. A quantitative trait, x, is assumed, which is influenced by a number of loci, as well as by the environment. The phenotypic variance of the variable x is σ_P^2. The number of loci and the influence of the environment are such that x follows a normal distribution, which we denote by $f(x)$ (for more on this distribution, see Background 4:3).

The selection on this phenotype is assumed to take place through truncation. This means that all individuals with a phenotypic value above a certain threshold, $x = T$, are selected to produce the next generations, whereas no individuals with a value below T are allowed to reproduce. Here, it is convenient to introduce the parameter Q_T, which stands for the area under the normal curve to the right of $x = T$.

When it comes to the considered locus A, we let its influence on x be such that the difference between the trait means of A_1A_1 and A_2A_2 individuals is $2a$ (again, a commonly used notation in quantitative genetics). To be able to mathematically handle the assumed selection scheme based on truncation, we choose to make three additional simplifying assumptions. The first is that the locus has only a small effect on the trait in question; the second is that there are many other loci affecting the trait; and the third is that the influence of the environment on the trait is substantial. Given this, it follows that $2a$ will be small relative to σ_P and that individuals with genotype A_1A_1 will have practically the same phenotypic variance, σ_P^2, as those with genotype A_2A_2. The trait distributions of the two kinds of homozygotes, which we denote $f_{11}(x)$ and $f_{22}(x)$, will thus be slightly shifted relative to each other but otherwise very similar. The trait distribution for the entire population, $f(x)$, will always fall close to and between these two distributions (which follows from our assumption of intermediary dominance). An illustration of the situation is given in Figure 9.1.

Combining the tools

We are now ready to tackle our primary aim, which is to find the relationship between S and s. It turns out that we can express both of these parameters in terms of the factor $\frac{f(T)}{Q_T}$, which stands for the value the normal distribution for the population takes at the truncation point

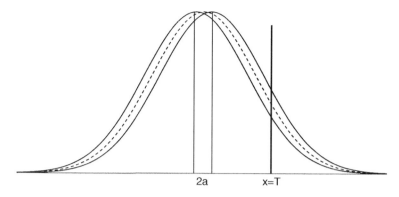

2a   x=T

Figure 9.1 Graphs showing (from left) the three distributions, $f_{22}(x)$, $f(x)$ (dashed) and $f_{11}(x)$, plus the threshold for selection given by T.

T divided by the area of the distribution to the right of this point. When this has been shown, the rest of the derivation follows easily. No more assumptions are made, but a mathematical approximation is needed, which is not particularly crude as long as a is small.

When truncation is made at point T, then this leaves a proportion that we denote Q_{T11} of A_1A_1 individuals surviving to reproduction, and another smaller proportion, Q_{T22}, of A_2A_2 individuals (remember that we have assumed that A_1 is the favoured allele). Since the variance is the same for A_1A_1 and A_2A_2 individuals (see our previous discussion about assumptions), their trait distributions, $f_{11}(x)$ and $f_{22}(x)$, will be identical, with the exception that their means are shifted $2a$ relative to each other, with the mean of A_1A_1 being greater than the mean of A_2A_2. For a graphical illustration, see Figure 9.1.

Since exactly the same area is found under $f_{22}(x)$ to the right of $T - 2a$ as under $f_{11}(x)$ to the right of T, the difference between Q_{T11} and Q_{T22} corresponds exactly to the hatched area in Figure 9.2. This area is closely approximated by $2a \cdot f(T)$, particularly for small values on a. Thus, we have that

$$Q_{T11} - Q_{T22} \approx 2a \cdot f(T), \qquad (9.3)$$

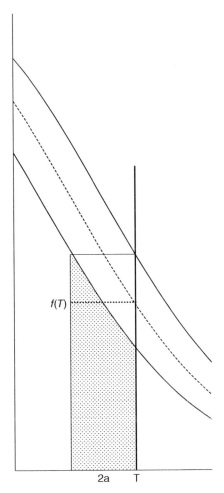

Figure 9.2 A close-up of that part of Figure 9.1 where the line $x = T$ cuts the three distributions $f_{22}(x), f(x)$ (dashed) and $f_{11}(x)$. The importance of the hatched area is described in the text.

which implies that

$$1 - \frac{Q_{T22}}{Q_{T11}} = \frac{2a \cdot f(T)}{Q_{T11}}.$$

Now, since Q_{T11} and Q_{T22} represent the proportions that survive to reproduction, they correspond directly to w_{11} and w_{22}, as introduced earlier. This implies that we have

$$\frac{w_{22}}{w_{11}} = 1 - s = \frac{Q_{T22}}{Q_{T11}}.$$

From these two results, it follows that

$$1 - s = \frac{Q_{T22}}{Q_{T11}} = 1 - \frac{2a \cdot f(T)}{Q_{T11}}$$

and that

$$s = \frac{2a \cdot f(T)}{Q_{T11}},$$

which again, using the fact that a is small, we can approximate as

$$s \approx \frac{2a \cdot f(T)}{Q_T}. \tag{9.4}$$

So much for the strength of selection at the considered locus, s; now we wish to know more about the strength of phenotypic selection, S. The selection that we envisage functions via truncation, and from Background 9:1 we have a nice result saying that truncating a normal distribution at point T increases its mean in such a way that

$$\mu_{Sel} - \mu_{Orig} = \frac{f(T)}{Q_T}\sigma_P{}^2.$$

Thus, according to our definition of the strength of the phenotypic selection, we have that

$$S = \frac{f(T)}{Q_T}\sigma_P{}^2. \tag{9.5}$$

It only remains for us to combine (9.5) and (9.4), which leads to

$$s = \frac{2a}{\sigma_P{}^2}S, \tag{9.6a}$$

Derivation 9

which also can be written as

$$s = 2\frac{a}{\sigma_p}\frac{S}{\sigma_p}. \tag{9.6b}$$

We have thereby found the relationship between the phenotypic measure of selection, S, and the genotypic measure of selection, s. The strength of selection at the locus in favour of allele A_1 is directly proportional to the strength of phenotypic selection, as well as the mean difference between the A_1A_1 and A_2A_2 individuals. Both these measures are in a natural way normalized by the trait's standard deviation due to genetic as well as environmental effects. (The result we have reached can be found in, for example, Robertson (1960), Crow and Kimura (1970) and Falconer and Mackay (1996), but in less explicit forms.)

Summing up

An interesting proportionality result has been reached, of the type that population genetics is good at producing. Here, the phenotypic strength of selection, which is measurable on some quantitative scale, is related to the strength of selection at a genetically variable locus, which may be measured as a surplus of genetic deaths. (An individual is considered genetically dead if it will not produce any offspring.) The two parameters are found to be directly proportional to each other with a proportionality constant given by how big the effect of the studied locus is relative to the variance of the phenotypic trait.

It is interesting to note that S, a and σ_P are all measured in the same unit as the trait. This implies that s is unitless – just as it should be, given that it measures fitness (which can be considered a probability value).

The exact degree of dominance at the studied locus is not important for the key result (the heterozygotes are intermediary and always close to the homozygotes). We return to the issue of the importance of the assumed mode of selection, and the independence of the contributing loci, later.

Interpretations, extensions and comments

The genetic effect of selection on a quantitative trait

Quantitative genetics stands on a solid Mendelian ground. It has done so since its start in 1918, when Fisher presented his analysis of the correlations to expect between relatives when part of the variation studied is due to mendelian genetic differences. In his derivation, he utilized the idea that many variable loci affect in an additive way what at the population level appears as a continuous trait. Empirical results supporting this possibility had been presented by, for example, our local predecessor Herman Nilsson-Ehle (1909) in his analyses of the seed colours in oats and wheat.

The relationship between quantitative genetics and population genetics has, on the other hand, always been less distinct. One reason for this is given in the introduction to this chapter: the whole point of quantitative genetics is to be able to ignore what happens at the many individual loci involved. A second reason is given by the result just derived: it implies that within the standard framework of quantitative genetics, the selection at the phenotypic level will have only small, often *very* small, effects at the individual loci – this is due to the necessary smallness of $\frac{a}{\sigma_p}$ in expression (9.6b).

Normally, strong effects have clear underlying causes. Here, however, we may have a strong effect – a substantial change in the mean of a trait due to selection – with an underlying cause that is only diffuse – the weak selective decrease in the frequency of disfavoured genetic variants at the many individual loci involved. This situation has a number of implications. The most obvious is that it will always be difficult to determine from studying a particular variable locus if it is involved or not in affecting a trait under selection. It also explains the fact, amply proven by experiments in Drosophila and various crops, that after one round of selection, the genetic variation for a selected trait normally remains almost unaffected, as it does after two, three, ten and one hundred rounds of selection; there may even still be a response to selection when the obtained trait mean falls standard deviations above or below the mean for the original population. The small effect each round of selection has on the variation at the participating loci is, obviously, the cause of this phenomenon.

Many other implications of the established relationship between s and S are also noteworthy. They provide the topics for the rest of this discussion.

We begin by noting that the result we have produced does not depend on the exact details of the selection process assumed. Kimura and Crow (1978) showed that weakening the strict all-or-none nature of the threshold selection procedure used in our model does not affect the result in any significant way: neither with respect to the strength of selection, S, nor to its expected response in the next generation, R.

Doing the selection in a *completely* different way may, however, have a clear effect on the selective response. We have assumed that selection acts independently at the different loci contributing to the trait. If, instead, we assume some kind of rank-order selection, whereby all individuals are ranked with respect to how many favoured or disfavoured alleles they carry and selected accordingly, then the response R of selection will be much stronger for any given number of selective genetic deaths – which is natural, since the efficiency of the genetic selection will then be at its highest.

Whether individual fitness is a quantitative trait that follows the logic of selection at independent loci or the logic of rank-order selection was much discussed in connection with the question of how much the mean fitness in a population is reduced if there is heterozygote advantage at many loci. We considered this topic in Derivation 2 and mentioned the debate that followed the provocative analysis of Lewontin and Hubby (1966), leading to an estimate of mean population fitness in humans of 10^{-46}. Rank-order selection for fitness would make this result much less extreme – not so many with beneficial genotypes at some loci would die or under-reproduce because of unfortunate combinations at other loci – which shows the strength and potentiality of this mode of selection. But, as Lewontin (1974a) points out in his description of the debate, there are problems with rank-order selection, too, primarily since it makes inbreeding much more deleterious than recorded, which renders it less believable. It is, anyway, most unlikely that this mode of selection will be applicable when more ordinary quantitative traits are considered, related for example to ecological adaptation (of, say, bird beak size or the optimal time for flowering in spring) or practical use (*e.g.* piglet weight or relative oil content in rape seed). For such traits, our approximate result (9.6) can be trusted to hold.

The limits of selection and the nature of 'good' alleles

An interesting use of the kind of analysis that we have presented here was made by Allan Robertson in 1960, when he combined it with Kimura's calculations of the probability of fixation of an advantageous allele (see Derivation 6).

Robertson's argument follows naturally from our result (9.6): if phenotypic selection takes place for a long time in a limited population, then not all alleles potentially favoured by selection will increase in frequency until fixation. Some of them will certainly be lost due to chance, and the probability of this is particularly great for those variants that are associated with only weak selective effects, s. Robertson introduced the notion of a *limit to selection*, by which he meant that selection applied over a long time to a population restricted in size (and with no mutation) will never reach its potential level. Instead, some of the advantageous alleles will be lost during the selection process, thereby decreasing – limiting – the effect ultimately obtained.

By joining the ideas discussed here with those in Derivation 6, Robertson showed that it is always favourable if selection is allowed to act on *large* populations, since genetic drift plays a proportionally smaller role there – a point to remember for all practical breeders. Limiting the variance in offspring numbers is also a good idea, for the same reason (see the discussion on effective population size in Derivation 4). There is, in addition, a possibility that if selection is allowed to act on separate lines run in parallel, then these lines may end up losing different good alleles, which makes it possible to obtain further progress if the lines later are intercrossed.

Robertson also pointed out how difficult it is to measure and analyse the effects of truly recessive mutations in quantitative genetics and breeding theory. If there are rare recessive alleles that potentially affect a selected character in a positive way, then their effects will neither be seen nor be useable for selection as long as they are rare (since homozygotes will be *very* rare and will neither be detected nor contribute noticeably to the measurable phenotypic variation). Such alleles are, of course, particularly likely candidates for being lost due to genetic drift.

[179]

A phenomenon similar to the unavoidable limit of selection is that all selective processes, particularly if they are strong, tend to favour alleles with major effects on the selected trait. Our result (9.6) may look self-evident, but its claim that selection acts *preferably* at loci with relatively strong effects is not so (this follows from s being directly proportional to a). This effect is problematic, since we expect mutations with stronger effects on a particular trait to have, on average, lower *general* fitness than variants with smaller effects – a suggestion first proposed by Fisher in 1930. Thus, there is a definite danger that if one achieves a good selection response for some important trait, then this goal has been achieved at the cost of ending up with a strain/line/breed/population with reduced viability or fertility, as has been observed in many breeding programs.

Threshold selection and disease liability

Another development of the quantitative genetics framework was presented by Douglas Falconer in 1965. He outlined how the tools of quantitative genetics can be applied to all-or-none diseases (you either have them or you don't) that show some heritable behaviour. The trick is to posit a basic physiological 'liability' to the disease, influenced by genes as well as environment, and then let the disease appear only in those above a certain cut-off liability value. From data on disease incidence in relatives, one may then calculate various parameters for the underlying quantitative liability trait. It is, for example, possible to estimate its heritability, as defined earlier in connection with expression (9.2b).

This procedure can in many cases feel practical and useful, in that it gives a numerical value for what generally is taken to be the genetic component of the disease. But it may at the same time be questioned – an estimated heritability value is, for example, hardly of any help in alleviating or understanding any disease. And the trait for which the estimate is made, the posited liability, is in such cases never seen, nor in some sense known to exist. Not even genetic counselling is in need of this methodology, since calculating the relevant probabilities for developing a disease can often be done directly from the available estimates of concordance between relatives, without taking a detour into heritability estimates.

Actually, medical genetics' interest in quantitative genetics comes not from a need for its full theoretical framework, but from an interest in one of its key components: the assumption that the variation in a trait is at least partly due to variation at a number of individual loci acting in a more or less additive manner.

While the last several decades' success in understanding monogenetic diseases has been almost complete, it has turned out to be surprisingly difficult to understand, analyse and find cures for the many common diseases in humans with less well-defined genetic components. Partly due to some early promising results concerning the HLA region, where it was found that certain HLA haplotypes are overrepresented in diseases such as diabetes, a belief became common that also other non-Mendelian diseases would show similar distinct genetic components. The hope was that genetic variation at a few easily studied loci – not just one or two, but perhaps five to ten – would explain the basis and occurrence of many, if not all such common diseases. This hope has, however, been found unrealistic.

From many different kinds of studies of the human genome, it has now become clear that none of the common diseases with distinct genetic components depends on the allelic composition of only a limited number of loci. Instead, risk-associated alleles are found at very many loci, each having only a small effect on the disease risk. Furthermore, these variants – taken together – explain only a small proportion of the observed familial clustering in the disease (Manolio *et al.*, 2009).

Quantitative genetics is not suited for causal analyses

The difficulties encountered by human genetics in its attempts to understand and cure partly heritable diseases can hardly be blamed on problems with the quantitative genetic assumptions used in the analyses. If anything, the reverse holds true. What two decades of genomic studies, in humans but also in Drosophila and many other species, have told us is that:

- Quantitative traits (such as height) generally depend on allelic variation at very many loci in addition to environmental effects.
- The effect of the alleles that contribute to the population trait variation is normally small (which contributes to a low repeatability of results).

- The relevant allelic variation may well map to sites outside well-defined structural genes, which makes their number potentially very high.
- The variation at the population level follows closely what is expected if the different loci act additively, even though pairwise epistatic interactions are commonly detected when directly looked for (Hill *et al.*, 2008).

These results have empirically reinforced the basis of quantitative genetics theory: its standard assumptions and approximations have been found to be sound. But they imply also that quantitative genetics is primarily a tool for analysing trait variation in populations, and not for the understanding of genetic causes (Lewontin, 1974b). Situations with *very* many contributing loci, each having allelic variants with *very* small effects, are *very* difficult to tackle when it comes to determining the interactions that underlie the genetic component in the development of the trait. Unfortunately.

Background 9:1 The truncated normal distribution

Assume a normally distributed random variable X with mean μ and variance σ^2. The frequency distribution of x is then

$$f(x) = \frac{1}{\sigma\sqrt{2\pi}} e^{-\frac{(x-\mu)^2}{2\sigma^2}}$$

for $-\infty < x < \infty$ (see Background 4:2).

Now consider a second random variable X_T, which is defined by cutting off, 'truncating', $f(x)$ at $x = T$ and retaining only the rightmost part. The frequency distribution of X_T is thus

$$f_{X_T}(x) = 0 \ \text{ for } \ x \le T$$

and

$$f_{X_T}(x) = \frac{1}{Q_T} \frac{1}{\sigma\sqrt{2\pi}} e^{-\frac{(x-\mu)^2}{2\sigma^2}}$$

for

$$T < x < \infty,$$

where

$$Q_T = \int_T^\infty \frac{1}{\sigma\sqrt{2\pi}} e^{-\frac{(x-\mu)^2}{2\sigma^2}} \, dx.$$

Q_T is the integral of the normal distribution from T to infinity, which is the same as saying that it is the area under the normal distribution from T to infinity. It is necessary for the function of the truncated normal distribution f_{X_T}, since it ensures that this function behaves as a true probability distribution (*i.e.* that its integral equals 1).

The mean of X_T we denote μ_T, and it is obviously greater than μ. We can be more specific than this, since it is possible to find a relatively 'nice' expression for the mean of the truncated normal distribution:

$$\mu_T = \mu + \frac{f(T)}{Q_T}\sigma^2. \tag{1}$$

Thus, truncating the normal distribution from T increases the mean by an amount that can be expressed in terms of the original normal distribution's

Derivation 9

variance, σ^2, the value of the normal distribution at T, $f(T)$, and the area of the normal distribution to the right of this point, Q_T.

To prove this assertion, we assume for simplicity that $\mu = 0$ (the exact point for μ does not affect the question we study). Then

$$\mu_T - \mu = \mu_T = \frac{1}{Q_T} \int_T^\infty x \frac{1}{\sigma\sqrt{2\pi}} e^{-\frac{x^2}{2\sigma^2}} \, dx,$$

which equals

$$\frac{1}{Q_T} \frac{1}{\sigma\sqrt{2\pi}} \int_T^\infty x e^{-\frac{x^2}{2\sigma^2}} \, dx = \frac{1}{Q_T} \frac{1}{\sigma\sqrt{2\pi}} \left[-\sigma^2 e^{-\frac{x^2}{2\sigma^2}} \right]_T^\infty$$

$$= \frac{1}{Q_T} \frac{1}{\sigma\sqrt{2\pi}} \left[0 - -\sigma^2 e^{-\frac{T^2}{2\sigma^2}} \right]$$

$$= \frac{\sigma^2}{Q_T} \frac{1}{\sigma\sqrt{2\pi}} e^{-\frac{T^2}{2\sigma^2}} = \frac{\sigma^2}{Q_T} f(T) = \frac{f(T)}{Q_T} \sigma^2.$$

The only nontrivial step in this derivation is finding the primitive function to $x e^{-\frac{x^2}{2\sigma^2}}$, which, actually, is not very difficult. The relevant function can also be looked up in books or on the net.

Questions

1. Two pure (*i.e.* completely homozygous) lines of a plant differ in height. The mean of the first line is 60 cm and the mean of the second line is 100 cm. The lines are crossed to produce an F_1 population, which then is self-fertilized to produce an F_2 population. The environmental effects are independent of the genetic effects and are such that $\sigma_E^2 = 8$ cm^2. The difference between the lines is due to k loci of equal effect, with the first line having only decreasing alleles and the second line having only increasing alleles. Assume that at all loci the alleles act additively, and that there is complete additivity and free segregation between loci. What are the trait means and variances in the F_1 and F_2 populations as a function of the number of loci? Partly for later purposes, summarize the values for σ_A^2, σ_G^2, σ_E^2, σ_P^2 and h^2 in the F_2 population!

2. A *very* large random-mating population is studied with respect to the same trait as in Question 1 (we may, thus, exclude drift effects). Assume that the increasing allele has frequency 0.4 at all loci in the starting population. Selection is then applied to the population until all frequencies have changed to 0.6. What is the difference between the means of the starting and finishing populations? Express this difference in terms of the standard deviation of the additive genetic variance in the starting population, σ_A.

3. Return to the F_2 population in Question 1 and let k be equal to 100. If selection is applied to this population so that the mean changes 0.1 standard deviations of the trait distribution and thereafter random mating plus offspring production takes place, what trait mean is expected in the next generation?

4. Continue with the assumptions from Questions 1 and 3. What is the selective effect, s, at the loci involved? And what is the expected change in allele frequency at these loci?

Derivation 10

Evolutionary genetic analysis of the sex ratio

Our last derivation is devoted to the problem of how genetic systems evolve. In elementary courses, one often talks about 'the Laws of Genetics'. And it is true – the genetic material often behaves in such a regular and ordered manner that it becomes natural to describe it in terms of natural laws. But then one must at the same time ask where these laws come from. Why is the gametic output of heterozygotes so close to 1 : 1? Why are mitochondria only transmitted via the mother in animals? And why are chloroplasts transmitted via female gametes in almost all angiosperms but in male gametes in many gymnosperms?

The answers to questions like these are not to be found in the molecular structure of DNA, nor in the exact cytological details of meiosis and gamete formation. The answers must be given in terms of evolution through some kind of natural selection. Thus, as in all other biological situations, we posit that the living world looks the way it does because those following some alternative inheritance patterns have, on average, left relatively fewer descendants.

The topic we have chosen for our discussion of how genetic systems evolve under selection is the sex ratio. Here, again, a strong regularity is found in nature: When a species has two separate sexes, males and females, their ratio is almost always close to 1 : 1, a ratio that holds near enough also in humans. The ubiquity of this ratio intrigued Darwin, but he could not figure out how natural selection could produce such a ratio, so he chose to defer the problem to the future (Darwin, 1871). Fisher (1930) thought he had found a satisfying solution to the question, when

Understanding Population Genetics, First Edition. Torbjörn Säll and Bengt O. Bengtsson.
© 2017 John Wiley & Sons Ltd. Published 2017 by John Wiley & Sons Ltd.

he argued that all random-mating populations would evolve towards a state where equal investments are made in the two sexes. He reached this conclusion by considering the number of grandchildren that matings with different propensities of producing sons and daughters would lead to, but he never presented any strict population genetic derivation of his result. This we will do here (simplifying but following closely the analysis made by Uyenoyama and Bengtsson, 1979). And by doing so, we also present a general method to approach complex evolutionary questions involving direct and indirect selection, a method that incorporates ideas from quantitative genetics and game theory as well as population genetics.

Analysis

Assumptions and notations

We study a diploid organism with separate males and females in a very large population (all drift effects are thus ignored). Sex can be determined in many ways in nature; here, to relate to the classical discussions of the sex ratio by Darwin and Fisher and many others, we assume that the mother can determine the sex of the offspring that she produces. (This may be done, to give but one example, by depositing eggs at different sites – given that the temperature during early development determines the sex of the young, as is the case in some reptiles.) Normal females produce offspring that with probability α turn into males and with probability $1 - \alpha$ turn into females. At the autosomal locus M, a rare allele, M_2, changes – 'modifies' – these probabilities so that females of genotype M_1M_2 produce males and females in ratio β to $1 - \beta$ instead of α to $1 - \alpha$. Mating is at random in the population.

We assume also that offspring of the two sexes may be unequally 'costly' to produce, and we let the cost of producing a son compared to a daughter be in proportion 1 to ϕ. All females have a constant and equal amount to invest in their offspring, implying that if one sex is more costly than the other then any modification of the sex ratio will alter the size of the litter produced.

Our task is to find the sex ratio towards which evolution will move. We study the question by investigating the conditions for an initial increase of the modifier allele M_2, which if it increases in frequency will change the sex ratio in the population. Basic to our analysis is Fisher's suggestion that evolution will move the sex ratio in the population towards the value where females invest equally in sons and daughters.

Finding the recursion equation system

As in the case of sib-altruism in Derivation 8, we need only consider the standard homozygote genotype M_1M_1 and the new heterozygote type M_1M_2, since the homozygotes M_2M_2 will be very rare. A new complication that arises in the present case is, however, that males and females with the same autosomal genotype will have different relative frequencies. Let the frequencies of the two genotypes M_1M_1 and M_1M_2 among females be x_{11} and x_{12}, and among males y_{11} and y_{12}. This gives us the following mating type/offspring table:

| Mating type | | Litter | Daughters | | Sons | |
(♀×♂)	Frequency	size	M_1M_1	M_1M_2	M_1M_1	M_1M_2
$M_1M_1\times M_1M_1$	$x_{11}\,y_{11}$	1	$1\ \alpha$	0	α	0
$M_1M_1\times M_1M_2$	$x_{11}\,y_{12}$	1	$\tfrac{1}{2}(1-\alpha)$	$\tfrac{1}{2}(1-\alpha)$	$\tfrac{1}{2}\alpha$	$\tfrac{1}{2}\alpha$
$M_1M_2\times M_1M_1$	$x_{12}\,y_{11}$	C	$\tfrac{1}{2}(1-\beta)$	$\tfrac{1}{2}(1-\beta)$	$\tfrac{1}{2}\beta$	$\tfrac{1}{2}\beta$

Just as in Derivation 8, we do not need to consider any additional mating types, since their frequencies will be too rare.

The litter size C for females producing males with frequency β, given the assumption made about constant overall investment, is found from the relationship

$$(1 - \alpha) \cdot \phi + \alpha \cdot 1 = C \cdot (1 - \beta) \cdot \phi + C \cdot \beta \cdot 1,$$

which leads to

$$C = \frac{\alpha + \phi - \alpha\phi}{\beta + \phi - \beta\phi}. \tag{10.1}$$

When ϕ equals 1 (*i.e.* when males are as costly to produce as females), we get − as expected − that C equals 1 and that all litters are of the same size, irrespective of what sex ratio they contain. It is trivially true that C

also always equals 1 for $\alpha = \beta$, since the new mutant allele M_2 has then no effect.

Let us also find the proportion of males suggested by Fisher, which we will designate A. He claimed that evolution would push the population towards a state where equal investments are made in the two sexes, which implies a frequency of males, A, given by $1 \cdot A = \phi \cdot (1 - A)$, which in turn implies:

$$A = \frac{\phi}{1 + \phi}. \tag{10.2}$$

For ϕ equals 1, this says that the expected frequency of males is $1/2$, corresponding to a sex ratio of $1 : 1$.

Now we return to the mating type/offspring table and write the resulting recursion equations:

$$x_{11}' = [(1 - \alpha)x_{11}y_{11} + 1/2(1 - \alpha)x_{11}y_{12} + C1/2(1 - \beta)x_{12}y_{11}]/$$
$$[(1 - \alpha)x_{11}y_{11} + (1 - \alpha)x_{11}y_{12} + C(1 - \beta)x_{12}y_{11}]$$

$$x_{12}' = [1/2(1 - \alpha)x_{11}y_{12} + C1/2(1 - \beta)x_{12}y_{11}]/$$
$$[(1 - \alpha)x_{11}y_{11} + (1 - \alpha)x_{11}y_{12} + C(1 - \beta)x_{12}y_{11}]$$

$$y_{11}' = [\alpha x_{11}y_{11} + 1/2\alpha x_{11}y_{12} + C1/2\beta x_{12}y_{11}]/$$
$$[\alpha x_{11}y_{11} + \alpha x_{11}y_{12} + C\beta x_{12}y_{11}]$$

$$y_{12}' = [1/2\alpha x_{11}y_{12} + C1/2\beta x_{12}y_{11}]/[\alpha x_{11}y_{11} + \alpha x_{11}y_{12} + C\beta x_{12}y_{11}]$$

Since the x and the y values denote relative frequencies among females and males, respectively, they have here been given different normalizing denominators. According to our assumptions, x_{12} and y_{12} are small, which implies that their squares as well as their products can be ignored. It follows also that since x_{22} and y_{22} are so very small, they can be ignored (and that therefore $x_{11} + x_{12} \approx 1$ and $y_{11} + y_{12} \approx 1$). The relevant information in the recursion system can therefore be reduced to and approximated by (see Background 2:2):

$$x_{12}' = [(1/2(1 - \alpha)x_{11}y_{12} + C1/2(1 - \beta)x_{12}y_{11}]/$$
$$[(1 - \alpha)x_{11}y_{11} + (1 - \alpha)x_{11}y_{12} + C(1 - \beta)x_{12}y_{11}] \approx$$

$$\approx [\frac{1}{2}(1 - \alpha)(1 - x_{12})y_{12} + C\frac{1}{2}(1 - \beta)x_{12}(1 - y_{12})]/$$

$$[(1-\alpha)(1-x_{12})(1-y_{12}) + (1-\alpha)(1-x_{12})y_{12} + C(1-\beta)x_{12}(1-y_{12})]$$

$$\approx [\frac{1}{2}(1 - \alpha)y_{12} + C\frac{1}{2}(1 - \beta)x_{12}]/$$

$$[(1 - \alpha) - (1 - \alpha)x_{12} - (1 - \alpha)y_{12} + (1 - \alpha)y_{12} + C(1 - \beta)x_{12}]$$

$$\approx [\frac{1}{2}(1 - \alpha)y_{12} + C\frac{1}{2}(1 - \beta)x_{12}]/(1 - \alpha)$$

$$y_{12}' = [\frac{1}{2}\alpha x_{11}y_{12} + C\frac{1}{2}\beta x_{12}y_{11}]/[\alpha x_{11}y_{11} + \alpha x_{11}y_{12} + C\beta x_{12}y_{11}]$$

$$\approx [\frac{1}{2}\alpha(1 - x_{12})y_{12} + C\frac{1}{2}\beta x_{12}(1 - y_{12})]/$$

$$[\alpha(1 - x_{12})(1 - y_{12}) + \alpha(1 - x_{12})y_{12} + C\beta x_{12}(1 - y_{12})]$$

$$\approx [\frac{1}{2}\alpha y_{12} + C\frac{1}{2}\beta x_{12}]/[\alpha - \alpha x_{12} - \alpha y_{12} + \alpha y_{12} + C\beta x_{12}]$$

$$\approx [\frac{1}{2}\alpha y_{12} + C\frac{1}{2}\beta x_{12}]/\alpha$$

Simplified one step extra, this becomes:

$$x_{12}' = \frac{1}{2}\left[\frac{(1 - \beta)C}{1 - \alpha}\right]x_{12} + \frac{1}{2}y_{12}$$

$$y_{12}' = \frac{1}{2}\left[\frac{\beta\, C}{\alpha}\right]x_{12} + \frac{1}{2}y_{12} \tag{10.3}$$

We have now reached the key recursive equations for our system. The situation with no variation, corresponding to $x_{12} = y_{12} = 0$, is obviously an equilibrium point. By looking at its stability, we determine under what condition the new, rare modifier allele M_2 will spread and thereby change the population's sex ratio.

Testing for stability

The recursion system is rather simple; with our assumption that the x_{12} and y_{12} values are small, we have been able to 'linearize' the relevant equations (see Background 6:1). In Background 10:1, we describe how the recursion system derived can be expressed in matrix notation and further analysed using linear algebra. The system (10.3) is, however, so simple that there is no need to use the matrix representation and fall back on an eigenvalue analysis. Simple addition of the two equations in (10.3) gives

$$x_{12}' + y_{12}' = \left[\frac{(1-\beta)C}{2(1-\alpha)} + \frac{\beta C}{2\alpha} \right] x_{12} + y_{12}, \qquad (10.4)$$

from which follows that if

$$\frac{(1-\beta)C}{2(1-\alpha)} + \frac{\beta C}{2\alpha} > 1 \qquad (10.5)$$

holds, then the new modifier allele M_2 will increase in frequency and the sex ratio will change. If, on the other hand, expression (10.5) does not hold, then the present sex ratio is stable to the perturbation and will not change.

Our only remaining task is therefore to insert the expression for C (which we derived earlier – see (10.1)) and simplify. This final step also illustrates one of the exciting properties of population genetics: Often, one runs into calculations that look terribly complicated and seem to lack any obvious structure, and then – suddenly – a number of factors and terms disappear, and a nice and crisp result appears out of the mess.

Thus, the present sex ratio is unstable if

$$\frac{(1-\beta)C}{2(1-\alpha)} + \frac{\beta C}{2\alpha} > 1 \iff$$

$$(1-\beta)\alpha C + \beta(1-\alpha)C > 2\alpha(1-\alpha) \iff$$

$$\alpha(1-\beta) + (1-\alpha)\beta > 2\alpha(1-\alpha)/C \iff$$

$$(\alpha + \phi - \alpha\phi)(\alpha + \beta - 2\alpha\beta) > (2\alpha - 2\alpha^2)(\beta + \phi - \beta\phi) \iff$$

$$\alpha^2 + \alpha\phi - \alpha^2\phi + \alpha\beta + \beta\phi - \alpha\beta\phi - 2\alpha^2\beta - 2\alpha\beta\phi + 2\alpha^2\beta\phi >$$

$$> 2\alpha\beta + 2\alpha\phi - 2\alpha\beta\phi - 2\alpha^2\beta - 2\alpha^2\phi + 2\alpha^2\beta\phi \iff$$

$$\alpha^2 - \alpha\phi + \alpha^2\phi > \alpha\beta - \beta\phi + \alpha\beta\phi \iff$$

$$\alpha[\alpha - \phi + \alpha\phi] > \beta[\alpha - \phi + \alpha\phi] \iff$$

$$\alpha[\alpha(1+\phi) - \phi] > \beta[\alpha(1+\phi) - \phi] \iff$$

$$\alpha \left[\alpha - \frac{\phi}{1 + \phi} \right] > \beta \left[\alpha - \frac{\phi}{1 + \phi} \right] \Longleftrightarrow$$

$$\alpha[\alpha - A] > \beta[\alpha - A] \Longleftrightarrow$$

$$\beta(A - \alpha) > \alpha(A - \alpha), \qquad (10.6)$$

where A, defined by (10.2), is the optimal frequency for male offspring as suggested by Fisher.

Let us now see what the instability condition

$$\beta(A - \alpha) > \alpha(A - \alpha) \qquad (10.7)$$

implies! (Observe that since $A - \alpha$ may be positive or negative, it cannot be just factored out of our expression without affecting the direction of the 'greater than' sign.) We start by assuming that the present frequency of males, α, is smaller than the suggested optimal vale, A. Then, the condition for spread of the new allele is that $\beta > \alpha$ (*i.e.* that the mutation leads to a higher population frequency of males). For $\beta < \alpha$, the mutation will not spread.

A similar argument for when α is greater than A shows that the new allele will then spread if $\beta < \alpha$ and not otherwise (*i.e.* that the new mutant leads to a smaller proportion of males in the population).

The case $\beta = \alpha$ is irrelevant in both cases, and there is no need to analyse it. For a graphic illustration of the final step in the analysis, see Figure 10.1.

Figure 10.1 Condition for the spread of a new modifier allele, M_2, changing the sex ratio produced by its female carriers. The optimal frequency of males, predicted by Fisher (1930), is denoted A, while the frequency of males currently produced in the population is α. In the illustrated case, α is smaller than A, which implies that for the modifier allele to spread, it should be associated with a frequency of sons, β, greater than α. A modifier associated with a smaller β value – in the figure, denoted β^* – will thus not spread.

Summing up

Thus, Fisher's suggestion is correct! The frequency of males will, under the outlined conditions, always move towards the state where there is an equal investment in the two sexes. When it is as costly for females to produce sons as it is to produce daughters, the commonly observed sex ratio 1 : 1 is therefore expected.

Interpretations, extensions and comments

Sex ratio selection

The result we have reached is baffling. In the assumed model, all the males participate equally in fertilizing the females and all the females produce offspring at their maximal capacity. No obvious source of selection is therefore at hand. Still, the sex ratio in the population will shift – if the conditions are right – due to a change in the frequency of alleles involved in determining the sex of the offspring. This evolutionary change can only be due to some kind of indirect selection that follows from the assumptions.

The result may verbally be explained as being due to the fact that matings involving allele M_2 produce potentially more offspring in the second and later generations than matings involving only the standard allele M_1. If M_2 has an advantage, it derives from the allele's help in producing a reproductively better mix of sons and daughters, given the prevalent sex ratio in the population. An allele that supports the production of the underrepresented sex will always carry an evolutionary advantage. Thus, we have seen that what can be called *indirect*, or *secondary*, *natural selection* may well act on a genetic system and change it – modify it – in some specific way. The most satisfying way to analyse such intricate processes is to put them into a well-defined population genetic framework, just as we have done in this derivation.

The close fit between the sex ratio commonly observed in nature and what our model predicts has an additional implication of wide general importance. It shows that the conditions necessary for group selection to function – discussed in Derivation 8 – do not normally prevail. This is directly obvious, since if group selection were able to determine the sex ratio, then a much more female-biased sex ratio would be expected (group selection always favours a high reproductive output from the groups involved, which gives an advantage to groups with many females). Since female-biased sex ratios are seen only rarely in nature (and the exceptions are normally better explained by other mechanisms – see some examples later in the chapter), it is reasonable to conclude that the selective forces acting inside populations are, in general, more important than the selective forces between-group competition gives rise to.

An explanation of well-delimited validity

What generally is called 'Fisher's theory of the sex ratio' should, perhaps, be seen not so much as a result within theoretical population genetics, but as a statement about what kind of variation exists that may influence the sex ratio. The theory says that if the sex ratio is modified by genetic variation which behaves in a Mendelian way and is inherited on autosomal chromosomes, then an equal investment in the two sexes is expected. Its usefulness comes from the fact that this background assumption very often holds. Very often, but not absolutely always. Occasionally, genetic variants of other kinds affecting sex determination occur, with subsequent effects on the sex ratio.

The easiest analysed case of such anomalous behaviour concerns 'X chromosome drivers': genetically different X-chromosomes that 'push for themselves' in meiosis and/or gametogenesis of XY males. They do so to the detriment of Y-carrying sperm, and they result in more XX daughters being produced than XY sons. Such variant chromosomes will spread due to the selective force that the segregation distortion confers, unless counteracted by other fitness components. At the population level, they lead, obviously, to a more female-biased sex ratio than normal. An early description of such a situation in *Drosophila obscura* was given by Gershenson (1928).

X-drivers are not the only possible type of selfish genetic elements that increase their presence in future generations by messing up the normally well-defined sex determination system and thereby changing a species' sex ratio. Other examples are given by Y-drivers, cytoplasms that always make their carriers into females, intracellular bacteria that do the same, and so on. The potential ways to produce odd sex ratios are many.

[195]

Fisher's sex ratio theory is, thus, not expected to hold always and every-where – in a population faced with an unusual mutation that in some more or less complicated way *directly* favours its own inheritance, the Fisher logic does not apply. It will, however, do so in a potential second step: Assume a population where an X-driver has become common – to continue with this example – and where more females than males therefore are produced. Any new variant allele that appears on an autosomal chromosome which in males restores the skewed segregation of the sex chromosomes towards a more normal value will then spread, following the Fisherian argument exactly as described in the derivation (though now the sex ratio in offspring is determined by the genotype of the father rather than the mother). Selection favouring new modifier alleles will obviously continue until the sex ratio in the population is back at 1 : 1.

Thus, the fact that some instances of 'extraordinary sex ratios' are found in nature (Hamilton, 1967 and others), does not mean that Fisher's analysis of the sex ratio question is wrong or irrelevant. When such instances of biased sex ratios have been carefully analysed, it has almost always been found that some strange sex ratio-controlling mutation is implicated in them, just like the X-drivers just discussed. Regarded from a wider per-spective, the predominance of the 1 : 1 sex ratio can therefore be seen as an indication that genetic variation inherited in a Mendelian way via autosomal chromosomes – and thereby fulfilling the assumptions nec-essary for making Fisher's sex ratio theory valid – predominates in the evolution of eukaryotic sexual organisms. Very loosely, one can say that a well-ordered machinery for inheritance is always in the interest of the large majority of the genetic material.

Actually, it is not only genetic variants behaving in a non-Mendelian manner that may invalidate the Fisher argument. Deviations from the posited random mating may also have an effect. If there is a high frequency of brother–sister matings – such as occurs, for example, in some parasitic wasps, where females lay their eggs in butterfly larvae and the young sibs mate before dispersing – then the conditions for Fisher's argument are not valid anymore, and a female-biased sex ratio is expected and often observed.

Yet another instance where non-random mating influences the distribution of resources to the two sexual functions is seen in self-fertilizing plants. There, plants with a high degree of selfing generally develop much smaller anthers and much fewer pollen grains than plants that cross-fertilize. Among the common West Asian/European grass cereals – barley, wheat, rye and oat – only rye is obligatorily wind-pollinating, and it invests, accordingly, much more in large anthers with many pollen grains (*i.e.* in the male function) than the other crops.

Meiotic recombination is an evolved genetic system

Actually, we have earlier in this book encountered another important case of selection on a genetic system. In the discussion of linkage disequilibria in Derivation 3, we described how the evolution of recombination – the regular breaking up of allelic combinations along chromosomes – can be explained. Keightley and Otto (2006) assumed an organism with many loci at which deleterious mutations occur, and considered how it would evolve in limited, but not necessarily small, populations. They were then able to show that genetic variants favouring increased recombination between the assumed loci would spread over time.

This is, again, an instance – like the sex ratio situation discussed here – where selection acts on the evolution of a genetic trait, recombination, in an altogether indirect way. No positive or negative fitness effects can be *directly* imputed to the genetic modifiers considered; selection acts on them only through their effects via recombination on other traits.

It is self-evident that the analysis of such indirect selective effects easily becomes very complicated. In the case of recombination, no reasonably simple analytical methods were available to Keightley and Otto to describe the dynamics in a succinct manner; they were therefore forced to use computer simulations instead. Sometimes, however, it is possible to take recourse to the elegant analytical procedure used here when studying complex evolutionary processes.

Evolutionary genetic analysis

In the present analysis of the sex ratio, and in the analysis of sib selection made in Derivation 8, we have seen what advantage may be gained from limiting the analysis to the behaviour of genetic variants when they are new and therefore rare. As long as random mating holds, one may then ignore the homozygotes for the considered variant. This methodology was developed for the evolutionary study of recombination rates by Nei (1967) and Feldman (1972).

It can be seen as a weakness that a full analysis of such assumed genetic situations cannot be made. In, for example, Derivation 2, all the equilibrium points to the recursion system were found and their stability properties formally demonstrated. This level of detail is, however, in many situations unnecessary, particularly if the question concerns some evolutionary property of more general importance. What one then wants to know is just the expected behaviour of a new genetic variant that modifies the trait of interest away from its currently standard value. Knowing the initial dynamics of an assumed modifier is perfectly sufficient.

It is, however, possible − and sometimes necessary − to simplify the population genetic analysis one step further. In our derivation, we tested a new allele, associated with effect β, in a population where all the rest of the genetic material caused effect α, and it turned out that the new variant would spread if it moved the population mean closer to the optimal value A. Performing such an initial increase analysis may, however, still be very difficult, and a helpful simplification − which retains the core of the logic in the analysis − is then to analyse only small changes to the prevalent population situation. As an example, if we only looked at the evolutionary fate of new mutations with effect $\alpha + \delta$, where δ is a small positive or negative number, rather than with the more general effect β, then this assumption of smallness would be simplifying, since all squared terms in δ^2 could then be ignored. (A more detailed description of the procedure is outlined by Christiansen, 1991.)

If we make this change to the analysis above, and substitute $\alpha + \delta$ for β, then the instability condition (10.7),

$$\beta(A - \alpha) > \alpha(A - \alpha)$$

becomes instead

$$(\alpha + \delta)(A - \alpha) > \alpha(A - \alpha),$$

Figure 10.2 Condition for spread of a new modifier allele, M_2, changing the sex ratio produced by its female carriers. Notations are as in Figure 10.1, though here only small perturbations, of size δ, of the currently produced frequency of males are considered. In the illustrated case, with α smaller than A, the modifier will spread given that δ is positive.

which simplifies to

$$\delta(A - \alpha) > 0.$$

Thus, if the proportion of males in the population is smaller than the suggested optimal value A, then the mutant will spread if it modifies the sex ratio by increasing the frequency of males; if the population frequency is greater than A, the mutation will spread if it decreases this frequency – just as expected. An illustration of the situation is given in Figure 10.2.

This method of analysing evolutionary questions combines a number of forceful ideas. First, it takes from quantitative genetics the analysis of alleles with only small effects on the trait in question. Second, it adopts from evolutionary game theory (Maynard Smith and Price, 1973) the suggestion that a given population situation should be tested with new types playing an alternative strategy. And it unites them within the well-structured framework of population genetics, where the representation of the different variants in coming generations is described with impeccable clarity.

The last point is particularly important when it comes to the evolution of genetic parameters, such as the segregation ratio in heterozygotes (normally unquestioningly assumed to be 1 : 1). Evolutionary game theory has here a distinct shortcoming in that it implicitly assumes a fair transfer of gains between generations (the one that wins in one generation transports its gains into a larger number of offspring in the next), which makes it less suited for analysing effects associated with 'unfair' systems of transmission. In such situations, it is always preferable to perform the analysis based on exact and explicit transmission descriptions – just as we have done in this chapter with respect to autosomal modifiers of the sex ratio.

Applied to interesting traits, this method for evolutionary genetic analysis gives a well-grounded feeling for how the more or less indirect selective forces on key genetic traits act. It provides the best available Darwinian method of analysis for how such traits are expected to evolve in nature.

Background 10:1 Stability analysis using linear algebra

Using vector and matrix notation, (10.3) can be written as:

$$\begin{pmatrix} x_{12}' \\ y_{12}' \end{pmatrix} = \begin{pmatrix} \dfrac{(1-\beta)C}{2(1-\alpha)} & \dfrac{1}{2} \\ \dfrac{\beta C}{2\alpha} & \dfrac{1}{2} \end{pmatrix} \begin{pmatrix} x_{12} \\ y_{12} \end{pmatrix}. \tag{1}$$

It is a standard result in linear algebra that the long-term behaviour of such a recursion system is given by the size of the largest *eigenvalue* of the 2 × 2 matrix. When this eigenvalue is numerically greater than 1, the modelled perturbations will increase with time, while they will decrease towards zero if the absolute value of the largest eigenvalue is smaller than 1. In the first case, the equilibrium under investigation is unstable, while in the second case the equilibrium is stable. It can be seen that condition (2.4) in the analysis of heterozygote advantage in Derivation 2 can be regarded as a one-dimensional version of the present expression (1).

For readers used to linear algebra, it is easy to find the characteristic polynomial, $CP(\lambda)$, of the matrix in (1). Its roots − representing the eigenvalues of the matrix - are given by

$$\lambda^2 - \lambda \left[\frac{1}{2} + \frac{(1-\beta)C}{2(1-\alpha)} \right] + \frac{(1-\beta)C}{4(1-\alpha)} - \frac{\beta C}{4\alpha} = 0.$$

The numerically largest of these roots falls at

$$\lambda = \frac{1}{4} + \frac{(1-\beta)C}{4(1-\alpha)} + \sqrt{\left[\frac{1}{4} + \frac{(1-\beta)C}{4(1-\alpha)} \right]^2 + \frac{\beta C}{4\alpha} - \frac{(1-\beta)C}{4(1-\alpha)}}.$$

This root is strictly greater than 1 if and only if

$$\frac{\beta C}{2\alpha} + \frac{(1-\beta)C}{2(1-\alpha)} > 1. \tag{2}$$

We find that condition (10.5) appears once more, just as expected.

In this particular case, the matrix notation did not decrease the amount of calculations needed. Rather, the reverse was true. However, in many

Derivation 10

other cases, studying the eigenvalues of the *stability matrix* is by far the easiest way to understand the behaviour of perturbations away from an equilibrium.

In practice, the size of the largest eigenvalue will often be known from just checking the characteristic polynomial evaluated at $\lambda = 1$. This is true for the present case, because $CP(\lambda = 1)$ corresponds to

$$1 - \frac{1}{2} - \frac{(1-\beta)C}{2(1-\alpha)} + \frac{(1-\beta)C}{4(1-\alpha)} - \frac{\beta C}{4\alpha} = \frac{1}{2} - \frac{(1-\beta)C}{4(1-\alpha)} - \frac{\beta C}{4\alpha}$$

$$= \frac{1}{2}\left[1 - \frac{(1-\beta)C}{2(1-\alpha)} - \frac{\beta C}{2\alpha}\right],$$

and the condition for (in)stability becomes the same as in (2).

Questions

1. *Evolution of apomixis.* Consider a population of hermaphroditic plants in which all individuals reproduce sexually via outcrossing. Assume that in this population, the normal wild-type allele, A_{sex}, mutates to a new dominant allele, A_{apo}, which makes its carriers apomictic (*i.e.* able to produce offspring via seed without any prior fertilization). These offspring have, thus, the same genotype as their mother. Since mother plants heterozygous for the A_{apo} allele produce all their seed without any fertilization, there can – obviously – be only two genotypes in the population: $A_{sex}A_{sex}$ plants, which produce seed after pollination and fertilization, and $A_{sex}A_{apo}$ plants, which produce seed apomictically, without any fertilizations.

   Assume also that the apomictic plants continue to produce as much pollen as normal plants, and let F be the amount of seed that apomict plants produce relative to sexual plants. (If F is 1, then the apomicts produce not only as much pollen as the sexual plants but also as much seed; if F is smaller than 1, then the mutational change in the breeding system has caused some decrease in the apomictic plants' seed-producing ability.) Then:

   (a) Set up a mating table with the different types of mother plants, the different types of pollen and the offspring that are produced.

   (b) Derive the recursion equation for the frequency of apomicts.

   (c) According to this equation, what happens when $F = 1$?

   (d) What happens for other values on F?

2. *Evolution of B chromosomes.* B chromosomes, also known as accessory chromosomes, have been found in many eukaryotes, such as crickets and grasses. The chromosomes lack functions that are necessary for their carriers, and they are often associated with self-promoting behaviour in meiosis and gamete production. (In an early article, Östergren (1945) talked about their 'parasitic nature'.)

   Consider an outbreeding animal species with an equal sex ratio. Individuals with no or one B chromosome have normal fitness, whereas individuals with two B chromosomes die before attaining reproductive age. Females with one B chromosome produce a proportion d of eggs carrying the B chromosome and a proportion

1 − *d* without the B chromosome (0 < *d* < 1). Males produce equal proportions of sperm with and without the B chromosome. Based on these assumptions:

(a) Set up the relevant mating table.

(b) Derive the recursion equation for the frequency of animals with one B chromosome.

(c) For what values on *d* will a new B chromosome enter and spread in the population?

(d) What frequency of animals carrying a B chromosome will there be in the population in the long run?

3. *Evolution of CMS*. In plants, mitochondrial mutations are known that cause their carriers not to produce any pollen – an effect called 'cytoplasmic male sterility' (CMS). The phenomenon is common in some angiosperm families, and it plays an important role in the breeding of many agricultural crops.

Consider an outbreeding plant species in which a mutation for CMS occurs and in which plants with this mutation produce an amount of seed that is *F* times the normal. (By not expending any energy on male function, seed output may in CMS plants be somewhat increased – in quality and/or in quantity – leading to *F* > 1.) The mutation for CMS is inherited strictly via the mother.

(a) Derive a recursion for the frequency of plants with the CMS cytoplasm and characterize its behaviour.

(b) What will happen in the long run to the population if *F* > 1, according to this recursion system?

(c) What scenario do *you* find realistic and likely for the population?

What's next?

With these ten derivations, our point has been made. We have shown how theoretical population genetics is done, what underlying assumptions are necessary and what results may be obtained. The complexity of life has been analytically broken down into questions of what will happen at a chromosomal locus with simple allelic variants. And we have described what knowledge about the functioning of life may be gained from this procedure. To continue in the same vein would be mechanical as well as superfluous. Readers who wish to learn other parts of population genetics should turn to other sources.

But if we had continued, which topics would we have covered? The following four are likely candidates. They concern important and relevant questions, and they contain striking results worth remembering. We present them via short introductions to round off our book.

Estimates and tests in population genetics

The world of population genetics has recently seen a major revolution. With new DNA technologies, genome sequence data have become available for all organisms in enormous quantities, at least in principle. With this development, the need for statistical treatments has become much more widespread. (Christiansen (2008) gives a valuable introduction to the field of molecular data and its population genetic analysis, primarily aimed at statisticians and bioinformaticists.)

While the mathematics that we use in this book is known to everybody studying natural science or medicine, no similar prior knowledge of statistics can be taken for granted. There are also distinct schools of

Understanding Population Genetics, First Edition. Torbjörn Säll and Bengt O. Bengtsson. © 2017 John Wiley & Sons Ltd. Published 2017 by John Wiley & Sons Ltd.

statistics that approach data and the inferences one may draw from them in different ways. Furthermore, to fruitfully discuss statistical questions related to real data, one must normally be more limited and precise (and not so general and abstract) than what we have been here. Statistical questions have therefore not been approached in this book. The interested reader is instead referred to the specialized treatises available (by *e.g.* Weir, 1996 and Balding *et al.*, 2007).

We wish, however, to make one comment concerning the relationship between population genetics and statistics: Almost all statistical methods used in this field are based on results coming out of theoretical population genetics. Anyone who wants to apply statistics to a population sample of some kind would therefore do well to first grasp the relevant theoretical background. Let us illustrate this with two examples, one concerning an estimate, the other a test.

The first concerns how the frequencies of haplotypes are estimated in a sample obtained from a diploid organism. Let us consider two DNA sites closely situated along a chromosome with alleles A_1 and A_2 and B_1 and B_2, respectively. If we assume that the two studied loci are highly variable and our sample from the population is big, then nine genotypes will be scored: $A_1 A_1 B_1 B_1$, $A_1 A_1 B_1 B_2$, $A_1 A_1 B_2 B_2$, $A_1 A_2 B_1 B_1$, $A_1 A_2 B_1 B_2$, $A_1 A_2 B_2 B_2$, $A_2 A_2 B_1 B_1$, $A_2 A_2 B_1 B_2$ and $A_2 A_2 B_2 B_2$. This information is, however, not sufficient for a straightforward counting of the numbers of different haplotypes, since among the double heterozygotes, $A_1 A_2 B_1 B_2$, there are those with haplophase $A_1 B_1 / A_2 B_2$ and those with haplophase $A_1 B_2 / A_2 B_1$.

Standard methods exist for making the required estimates without knowing the exact linkage phase of the double heterozygotes, so the problem that we have outlined is not insoluble. But – and here comes our point – these estimates become *better* if the population is known to be random-mating; that is, if one may assume that all loci show Hardy–Weinberg proportions, including those in the sample. Thus, for the simple task of obtaining good estimates of the frequencies of the four haplotypes $A_1 B_1$, $A_1 B_2$, $A_2 B_1$ and $A_2 B_2$, and through them the degree of linkage disequilibrium between the loci, it is essential to have a good prior understanding of what random mating means and what Hardy–Weinberg proportions stand for.

Our second example concerns statistical methods for analysing deviations from neutrality and randomness, illustrated by the test introduced by Tajima in 1989 and discussed in Derivation 5. Again, we leave the details to others (see *e.g.* Felsenstein, 2004; Gillespie, 2004; Hedrick, 20011a; Hein *et al.*, 2005), and here simply stress the population genetic prerequisites necessary for understanding this standard procedure.

We start with a sample of homologous DNA sequences from a population. At some sites, there will be two bases, both at substantial frequencies, while other sites will be monomorphic. The processes leading to such variability have been discussed in Derivation 5; they are conveniently studied with the infinite sites model of mutations and the Wright–Fisher model for reproduction. When these two models are combined, many outcomes are possible – but not all. For example, we do not expect all sites to have the same level of substantial variation, just as we do not expect the variation to be high at a very small number of sites and nil everywhere else. Tajima's test is designed to check whether the variation in the sample shows an acceptable pattern somewhere between these two alternatives. When the test indicates a deviation from the null model given by the neutral evolutionary assumptions, there are reasons to conclude that either the mutation model or the population model does not hold, just as described in Derivation 5. Thus, we believe that it is next to impossible to give a biologically meaningful interpretation of a significant Tajima test value, if one has not first – at least once – worked through and grasped the assumptions of its population genetic background.

The mutation–selection balance

Life is by necessity a balance between mutation and selection. Wherever genetic information exists, deleterious mutations must occur – this follows from the laws of thermodynamics. And wherever life exists, natural selection counteracts the loss of adaptation due to such mutations. If the rate of mutation is above a critical limit, selection will not be able to counterbalance its deleterious effects, and life disappears.

This balance between mutation and selection has not been chosen as one of our topics of derivation; most textbooks in genetics cover the question well. The commonly taught results are the following: Consider deleterious mutations, A_0, that appear with frequency μ at a locus with standard allele A_1 in a diploid random-mating species (thus, we here distinguish between the standard allele, A_1, and a whole class of deleterious mutations regarded as A_0). The three possible genotypes $A_1 A_1$, $A_1 A_0$ and $A_0 A_0$ have relative fitness values 1, $1 - hs$ and $1 - s$. The population frequency of the deleterious alleles is then expected to be $\frac{\mu}{hs}$ if the mutations have a (partially) dominant deleterious effect (*i.e.* if $h > 0$), and $\sqrt{\frac{\mu}{s}}$ if the deleterious mutations are strictly recessive ($h = 0$). These results were first given by Haldane (1927).

However, this balance will not only be determined by mutation and selection in natural situations – genetic drift will also play a role. Wright (1937) gave formulae for the distributions of deleterious alleles under these combined evolutionary forces. Of particular interest is the fact that while the frequency of dominant deleterious mutations will be fairly close to its expected value, recessive deleterious mutations will show a much larger spread in frequencies. This result is important in human genetics, since it says that if one studies recessive deleterious mutations at many loci in a human population, then these mutations will be very rare at some loci (the associated disease will perhaps never be seen) and surprisingly common at others. Thus, if in a population a certain recessive genetic disease is found to be unusually frequent, it is difficult to know – without extensive further studies – whether this result is due purely to genetic drift of the underlying deleterious alleles or whether, for example, the heterozygote carriers of the alleles are associated with some small positive fitness effect.

Let us finish this section with an important observation due to Haldane (1927), which we will illustrate with recessive mutations. Many genetic diseases are caused by such mutations, but they may differ widely in their fitness effects (*i.e.* in their selection values, s) – some diseases are lethal, while others decrease fitness by only a small extent. Surprisingly, their effects on the mean fitness in the population are nevertheless the same, if they have the same mutation rate.

This result is seen as follows: If we use the model just discussed, then the frequencies of the three genotypes $A_1 A_1$, $A_1 A_2$ and $A_2 A_2$ are expected to be $(1 - q_{eq})^2$, $2(1 - q_{eq})q_{eq}$ and q_{eq}^2, with respective fitness values 1, 1 and $1 - s$. Given that $q_{eq} = \sqrt{\frac{\mu}{s}}$, the mean fitness in the population is expected to be:

$$
\begin{aligned}
W_{eq} &= (1 - q_{eq})^2 \cdot 1 + 2(1 - q_{eq})q_{eq} \cdot 1 + q_{eq}^2 \cdot (1 - s) \\
&= (1 - q_{eq})^2 + 2(1 - q_{eq})q_{eq} + q_{eq}^2 - q_{eq}^2 s \\
&= [1 - q_{eq} + q_{eq}]^2 - q_{eq}^2 s = 1 - q_{eq}^2 s = 1 - \sqrt{\frac{\mu}{s}}^2 \cdot s = 1 - \mu.
\end{aligned}
$$

The loss of fitness to the population due to the deleterious mutations is, thus, equal to the mutation rate, μ, and independent of whether the recessive mutants are associated with strong or mild effects. This loss of mean fitness is called the *mutation load*, and why this load does not depend on the strength of selection can be understood by realizing that all deleterious mutations must – sooner or later – be lost from a population, whether this occurs drastically with, say, the death of an infant or over many generations via a minute decrease in fertility. A similar argument holds for the aggregated fitness decrease of a population due to dominant deleterious mutations at a locus, but there the load is 2μ, since every selective removal of a deleterious allele takes with it one standard gene copy.

Thus, when deleterious mutations occur at all important loci in the genome, they inflict a fitness decrease to the population directly related to the mutation rate. This implies that there is a constant, secondary selection going on in the population against the induction of mutations. It is possible to use the modifier methods discussed in Derivation 10 to follow how otherwise neutral alleles that – directly or indirectly – decrease the mutation rate will increase in frequency. For modifiers that only affect a single locus, the force of this secondary selection will be very weak and not of any particular importance. Modifiers that decrease the mutation rates at all of the loci in a genome will, on the other hand, be associated with a strong selective advantage. The reason for the build-up of powerful

DNA repair mechanisms in all species with large genomes obviously has its roots here.

Partial genetic isolation

Among all the genomic information and knowledge that has been harvested from natural populations in recent years, it is not easy to pick out what is of particular theoretical importance. We would, however, claim that the finding of partially speciated genomes is of very special interest.

The phenomenon of 'partial speciation' had been discussed and analysed before, but a number of genomic studies of some critical evolutionary relationships in birds have now shown the general relevance of this phenomenon. When morphologically distinct but closely related evolutionary units are investigated – be they called 'species', 'subspecies' or 'races' – it is found that their genomes switch abruptly between parts that are similar and parts that are distinctly differentiated. Between the pied and the collared flycatcher, for example, there are many such diverged genomic stretches (Ellegren *et al.*, 2012), while between the black carrion crow and the grey-coated hooded crow, one major differentiated genomic stretch dominates (Poelstra *et al.*, 2014).

Let us use the crow situation for our discussion. Here are two evolutionary units that still form fertile hybrids. Individuals that are heterozygous for the differentiated chromosome region will, though, form fewer offspring than others – this is why the two types of crows don't become completely mixed despite the hybridizations. Assume that selection acting against such hybrid heterozygotes is of strength s. Other parts of the genome will not be directly affected by this selection and can, in principle, mix freely. The flow of the undifferentiated genes between the two forms will, however, be reduced by the partial divergence of the genomes. There is what can be considered a genetic barrier between the two forms.

How strong will such a barrier to gene flow be? It is clear that if a genetic barrier between two forms is very strong, then they will soon be genetically totally separated from each other and will constitute different species. If, on the other hand, the genetic barrier is weak, then the two forms will continue to evolve as a single evolutionary unit, except for that part of the genome which separates them as distinct.

This situation cries out for a population genetic analysis. Through the use of a very simple model, it can be shown that the genetic barrier of strength s will decrease the 'effective flow' of genes between the two forms by a fraction $2s/(1 + s)$ (Bengtsson, 1985; see also Barton and Bengtsson, 1986). This result holds for all those genes that recombine freely relative to the chromosome stretch that is differentiated. Should the two forms instead be genetically differentiated at many independent sites acting in a multiplicative manner, but with the fitness of first-generation hybrids still reduced to $1 - s$, then the result will be very similar, since the decrease in the effective gene flow becomes $2s(1 - s)$, which for small values on s is approximately the same as $2s/(1 + s)$.

From these results, we learn that if a chromosome region has evolved into two versions – perhaps helped by an inversion that prevents recombination within the region – that are well adapted separately but produce a lower fitness when together in heterozygous form, then a species may well split into two distinct units with a genetic 'tension zone' between them, given that matings are reasonably local. Should the heterozygotes for the differentiated region have a clear fitness disadvantage, of say 10%, then the zone between the two forms will be very stable, and the situation will from the outside look as if there are two distinct evolutionary units. However, the flow of unlinked genes between these units will be reduced by less than 20% compared to the value expected in the absence of the genetic barrier.

The way that partial genetic differentiation between closely related evolutionary units increases or decreases over time will probably become an important part of population genetics in the years to come. And it will – we believe – lead to many empirical and theoretical studies of great interest in showing a variable fluidity of the genetic material.

Segregation distortion and genetic conflicts

We introduced this book with a verbal description of the Law of Constancy of Allele Frequencies. It is natural to return to it now, and its apparent simplicity.

This law provides the living world with a canvas on to which evolution may paint amazing forms. The regular genetic system is in some sense evolutionarily inert, which implies that evolutionary forces can produce great results even with small effects (though it may take long times). This inertness also underlies a theme that we have not particularly stressed but which has been important in the history of population genetics, namely that the adaptation of organisms to their environments will irrevocably increase as long as evolution follows the rules of genetics. In Derivation 2, we described how selection acting on one locus with many alleles having fixed fitness values never decreases the mean fitness in the population, but otherwise we have been restrictive with this idea (which has led to many discussions and conflicts around Fisher's and Wright's different conceptions of evolution).

There are, indeed, good reasons for not letting population genetics concentrate on (only) explaining adaptation in nature, since many instances are known where organisms *do not* follow the rules of genetics. Or, better expressed: where organisms contain genetic material that disturbs the normal rules of Mendelian transmittance, so that this material becomes relatively overrepresented in later generations. A very particular example is given by transposing elements, which, by multiplying and inserting themselves at new positions in the genome, do not follow the rules of Mendelian inheritance – their population genetics is worth a book of its own. Another example is given in Derivation 10, where the X-drive in *Drosophila obscura* is described. Many other such non-Mendelian genetic factors are known, such as B chromosomes in plants (Östergren, 1945) and *t*-alleles in mice (Dunn, 1953). More uncertain, but highly interesting, is whether the very weak but ubiquitous effect of biased gene conversion that skews the Mendelian segregation ratio should be ascribed any major influence on evolution (Marais, 2003).

When adaption to natural conditions is considered, we – in our mind's eye – normally think of the reactions and behaviour of some diploid species. It then becomes obvious that all genetic factors which break the Mendelian rule of 1 : 1 segregation are potential threats to the adaptation of the organism – because a factor that pushes for itself during the haploid phase of transmission may well be associated with negative effects at the diploid level and still spread, as shown, for example, by *t*-alleles and many B chromosomes. When such factors spread in a population, the

[212]

diploid population's mean fitness – its degree of adaptation – obviously decreases.

An accurate way to describe such instances of dysadaptive evolution is to say that 'shit happens', in organismic life as anywhere else. But the scenario we have described does not necessarily have an unhappy ending. Genetic variation at other positions in the genome may influence the behaviour of the problematic genetic factor by either decreasing its deleterious effect or reducing its selfish transmission. Unless associated with substantial negative fitness effects of their own, such restorers will spread in the species and make it slowly return towards its original level of adaptation (the evolution of such restoring modifiers was discussed at the end of Derivation 10).

We have now entered the fascinating field of *genetic conflicts*, where one considers how strict evolutionary dynamics may lead a species into different – sometimes contradictory – phenotypic states, depending on what genetic variation it has to act on, and in particular how this variation is transmitted (for more on this phenomenon, see Hurst *et al.*, 1996 and Burt and Trivers, 2008). Many organisms even have what can be considered a constant inbuilt potential genetic conflict. This is the case for diploid sexual eukaryotes with maternal transmission of their mitochondria: From the point of view of this mitochondrial DNA, it is a waste of evolutionary effort to produce males (gametes, parts or whole organisms), since the DNA will not be further transmitted from there. Various genetic conflicts related to this effect have been detected; for example, the lack of pollen production in some angiosperms (Lewis, 1941; see Question 3 in Derivation 10). As expected, such drastic changes will directly lead to autosomal restorers being selectively favoured, and such factors are also commonly found in affected species.

All deeper analysis of DNA data requires relevant population genetics theory – this much is obvious. But population genetics theory cannot be dispensed with in evolutionary studies either, as illustrated by the examples just given of what may be seen as oddities, but which nevertheless are evolutionary realities.

It is true that Darwin could present an impressive argument for evolution via natural selection without any detailed genetic knowledge.

And – let there be no doubt about it – many aspects of organisms' interactions with their environment do not require any detailed genetic information, much less any formal population genetic modelling, to be understood. Various fitness-maximization techniques, assuming the availability of some suitable underlying quantitative genetic variation, are often perfectly sufficient to elucidate the underlying evolutionary causations. Likewise, when it comes to the evolution of behaviour, where fitness values are not distinct but depend on what others do, various game theory methods work well and explicit population genetics is not needed.

But when inbreeding starts to affect the composition of the offspring, or the transmission of genetic factors is unfair, then finding a good population genetic model which – even if crudely – incorporates all the relevant factors into an analytically tractable form becomes necessary. And very intellectually satisfying.

Epilogue

The natural path forward for readers who have worked their way through this book is to go back. Back to the beginning.

One learns population genetics by working through its basic results over and over and over again. The first time, it is a question of how it all works. The next time, it may be to see how different parts of the theory hang together and how different results interrelate. Later, the more advanced parts fascinate; at other times, it is the simplicity of the theory. As usual, the best way to learn something is to teach it. Over the years, we have many times seen how our graduate students gain a deeper understanding of the field by having to teach population genetics in practicals.

It is also good to have to apply or extend the theory. Anyone who deals with data from real biological material will sometimes question whether it is correct to use some standard population genetic tool or result, because the underlying assumptions are not exactly fulfilled. The best way to proceed is then to try to get a grasp of how important the deviant effect may be. Sometimes, a simple calculation is sufficient to tell what is reasonable and what is not. At other times, it is better or more convenient to simulate the problem in a computer.

Theoretical population genetics does not need to be more technically complex than shown here. We recommend, however, that anyone who wants to continue with further studies in the field learns matrix theory and linear algebra. But, in general, it holds that the mathematics often does not matter very much in population genetics unless one is particularly interested in the maths for its own sake. The importance is grasping the field's logical coherence, so that it is remembered when all the mathematical details have been forgotten.

Understanding Population Genetics, First Edition. Torbjörn Säll and Bengt O. Bengtsson.
© 2017 John Wiley & Sons Ltd. Published 2017 by John Wiley & Sons Ltd.

For many of the readers of this book, it will be natural to continue to learn more about population genetics within the framework of bioinformatics or evolutionary theory. Plenty of excellent books on these topics are available, and some can be found in our list of references. Due to the field's theoretical nature, the results of population genetics age very well. It is therefore often interesting and relevant to study also old texts, even if the notations used and the way the problems are formulated may at first appear strange or antiquated.

We hope, however, that some of the spirit in which this book has been written stays with all our readers, even those for whom population genetics will function mainly as a practical tool. Population genetics is for us where the logic of life becomes penetrable and unified. Population genetics is where Darwin's theory, ecological insights, Mendel's rules, and the molecular structure of DNA, all come together and are moulded into a coherent logical structure.

If one wants to *understand* life, in all its improbable and amazing richness, one must begin by understanding population genetics.

Thanks

We are grateful for financial support from the Nilsson-Ehle Endowments, administered by the Royal Physiographic Society of Lund. The staff at the Biology Library in Lund have been most helpful with our many queries.

We also acknowledge excellent professional contacts with Wiley and thank, in particular, the anonymous reviewer, who carefully went through the first version of the text, and our copy editor, Tim West, for invaluable help and advice.

Warm thanks goes to our families: Cornelia and Yvonne, Sara and Boel. Thanks also to our colleagues with whom we have discussed population genetic questions over the years. Associate Professor Peter Pettersson read the derivations in an earlier version and we are very grateful for his comments. The remaining mistakes are our own.

Because this is the bitter fact: In a book like this, it is impossible to avoid mistakes, typos and unsuitable wordings. There may even be some serious blunders in the text, despite our attempt to avoid them. We are highly grateful to everyone who contacts us to point out these weaknesses.

Our most deep-felt thanks goes to the students of the course 'Understanding Population Genetics', which has been given at Beijing Normal University and several times at the GENECO Research School at Lund University. Their encouragement provides the book's *raison d'être*.

Understanding Population Genetics, First Edition. Torbjörn Säll and Bengt O. Bengtsson.
© 2017 John Wiley & Sons Ltd. Published 2017 by John Wiley & Sons Ltd.

Glossary

Scientific words and concepts change meanings over time. There are also differences in usage between authors and intellectual traditions. Science will, however, only function if it is possible to understand – with a reasonable degree of accuracy – what others wish to convey. We hope that the following list of explanations will help the reader to grasp the meanings that we attach to a number of important terms in genetics and mathematics. Some elementary words are included to help readers with a background outside these fields. We have not striven for strict formal definitions, but have instead tried to give helpful descriptions of how the words are used. For words not included here, particularly in statistics and evolutionary theory, see the Index, which may refer to specific pages where they are introduced and discussed.

Additive effect – An allele acts additively if carrying one copy of it adds (detracts) half the value to a studied trait compared to carrying two copies. (This notion is, of course, valid only for diploids.) If variation at two or more loci affect a trait and the contributions of the loci function independently of one another (*i.e.* there is no epistasis), then one also talks about additive effects.

Allele – A distinct and recognizable type of a gene. Sometimes, the word is used as a synonym for what here is called 'gene copy'. Note that in this book we use 'gene' to stand for any sequence occupying a locus, not just those sequences that code for proteins or are associated with specific functions.

Allotetraploid – An individual from an allotetraploid species carries two haploid genomes originating from one species and two haploid

Understanding Population Genetics, First Edition. Torbjörn Säll and Bengt O. Bengtsson.
© 2017 John Wiley & Sons Ltd. Published 2017 by John Wiley & Sons Ltd.

genomes originating from another species. The phenomenon is common among plants.

Asexuality – When reproduction occurs without fertilization and meiosis. Under asexuality, the offspring have the same genotype as the single parent. In a partially asexual species, some offspring are produced via sexual reproduction while others are produced via some asexual method. 'Apomixis' is the term used for plants that produce seeds asexually; interestingly, such plants may still be able to produce pollen via sexual means (see Question 1 in Derivation 10).

Association study – An investigation that aims at finding associations – correlations – between traits (often diseases) and genetic variation somewhere in the genome. Such studies can investigate a set of chosen candidate genes, or they may be genome-wide association studies (GWAS), where the variation in the whole genome is scored via densely situated markers. The logic behind the latter kind of study is that either the marker variants are directly involved in the causation of the trait or they are in linkage disequilibrium with allelic variants at closely situated loci that are involved in the causation of the trait.

Autosomes – Chromosomes carrying genetic material not inherited in a sex-specific manner. The sex chromosomes are, thus, not autosomes. Diploid organisms have two copies of all autosomal chromosomes.

Boundary condition – The solution to a differential equation is normally a set of functions that differ with respect to one or more parameters (constants). In many situations it is, however, known what value the looked-for function should take for a specific value of the independent variable; this, then, gives the boundary condition from which the valid solution to the differential equation can be found. Example: The differential equation $\frac{dy}{dx} = y(x)$ is solved by all equations $y(x) = e^x + C$, where C is a constant. With boundary condition $y(0) = 1$, the only valid solution to the differential equation becomes $y(x) = e^x$.

Chloroplast – A cellular organelle that performs photosynthesis in, for example, plants and algae. Chloroplasts contain DNA and are in angiosperms normally transmitted via the egg cell.

Chromosome – A DNA molecule carrying genetic information. A chromosome undergoes orderly replication and is transmitted at cell division (in eukaryotes: mitosis) and reproduction (in sexual eukaryotes: meiosis). A eukaryotic chromosome as normally illustrated is a complex structure with a characteristic shape: the DNA molecule running from one end of the chromosome to the other is heavily coiled and condensed, and it is bound to various proteins. The DNA molecule has just been replicated and the two duplicated copies lie close to each other. Soon, these two copies will separate and help form two cells out of what originally was only one.

Coalescence – Two homologous gene copies must be derived from a single, earlier gene copy through a series of replications. This holds true whether the two copies are taken from a single individual or from two individuals belonging to widely different species. If the histories of these gene copies are followed back in time – which normally is only possible 'in principle' and without exact knowledge – then the moment when they come together is called the 'coalescence event'. If the size and the reproductive behaviour of a population is known, then the mean time to calescence (and other similar properties) of two randomly drawn gene copies can be estimated from the effective size of the population (as is done in Derivation 4). With the arrival of DNA sequence information, coalescence analysis has become a very important part of population genetics.

Descent, homozygous by descent, identical by descent Two gene copies are identical by descent if they both derive from a gene copy carried by a joint ancestor. If the two identical copies exist in the same individual, then this individual is homozygous by descent for the studied gene. Estimates from pedigree information of the probability of gene copies being identical constitute the basis for measures of relatedness and inbreeding.

Difference equation – An equation describing how a function changes over time, measured in discrete steps – in population genetics, normally generations. Almost all of the derivations in this book are based on analyses of difference equations.

Differential equation – An equation describing how a function changes over continuously running time, based on the derivative of the function. See Derivation 1, and particularly Background 1:2.

Diffusion process – A mathematically well-defined concept in physics that has been successfully employed to model random genetic drift in population genetics. If a neutral allele has frequency x in the current generation, then its frequency in the next is best represented by a probability distribution centred on x (or nearby, if the allele is affected by *e.g.* selection) and with a variance given by the strength of drift in the population (*i.e.* its effective population size). Derivation 6 gives a detailed example of such a use of diffusion mathematics in genetics. Another importance source of diffusion in population genetics, though not one studied here, is the geographic movement of gene copies in time over one- or two-dimensional space.

Diploid – An organism is diploid if it develops out of a cell containing two haploid genomes. All autosomal genes exist in two copies in a diploid individual.

Diversity, often *genetic diversity* or (*expected*) *heterozygosity* – A commonly used measure of the genetic variation in a population. The genetic diversity at a particular locus in a population, normally designated H, is the probability that two randomly drawn gene copies from the locus are of different allelic types. In a diploid population produced by random mating, the genetic diversity is equal to the frequency with which heterozygotes are formed; hence, the commonly used name 'heterozygosity' for this measure. H is, however, equally often used to measure the genetic diversity in haploid organisms, for which the word 'heterozygosity' does not make any sense.

Dominance – In Mendelian genetics, an allele A is dominant over a different allele a (called recessive) if the heterozygote Aa has the same phenotype as AA and this phenotype is different from the phenotype of aa. In quantitative genetics, dominance refers to the position of the heterozygotes relative to the two kinds of homozygotes on the scale by which the trait of interest is measured. In Derivation 9, for example, we let h be our measure of the degree of dominance with respect to fitness, at a locus where the homozygotes take values 1 and $1 - s$, respectively, while the heterozygotes take value $1 - hs$.

Drift, often *genetic drift* – The effect of 'pure chance' on the transmission of genetic variation from one generation to the next. Even when no evolutionary forces are present, the Law of the Constancy of Allele Frequencies cannot guarantee that the allele frequencies remain exactly the same in two consecutive generations, only that their *expected* frequencies are unaltered – this is due to the effect of random drift that comes with the discrete and finite nature of the genetic material.

Dynamic process, dynamic system – A process that describes the changes in a system over time. In population genetics, it is normally a set of allele or haplotype frequencies that is studied.

Effective population size or *number* – The size a Wright–Fisher population (see later) should have in order to show the same drift properties (mean time to coalescence and/or rate of loss of genetic variation in the absence of mutation) as a particular population under study.

Equilibrium – A state from which a dynamic system will not, in principle, move. If the system evolves away from this state when slightly perturbed, then the equilibrium is said to be unstable, while the equilibrium is locally stable if a small perturbation away from it decreases with time. The equilibrium is globally stable if the system always moves towards it, irrespective of the starting position of the system. At a boundary equilibrium, one of the possible types is missing; testing for its local stability thus implies investigating whether this missing type will spread when rare. An internal equilibrium is an equilibrium where all of the possible types occur.

Eukaryote – All known cellular life on earth belong to one of three groups: the bacteria, the archaea (these two together make up the prokaryotes) and the eukaryotes. In the eukaryotes, the genetic material (outside mitochondria and chloroplasts) is organised into a well-separated cellular organelle, the nucleus. Most eukaryotes reproduce sexually and regularly go through the meiotic cycle. A wide taxonomic range of eukaryotes are unicellular and live in water; among the larger multicellular eukaryotes, one finds fungi, plants and animals (including humans).

Fixation – The process by which a genetic variant, often an allele, becomes more and more common ('it spreads') in a population until all alternative variants have disappeared. At this moment, the allele is said to have become fixed or, alternatively, to have gone to fixation.

Frequency-dependent selection – This occurs when the fitness values of different genotypes are not constant, but depend on one another's population frequencies. The phenomenon is discussed in Derivation 9.

Gamete – A cell specialized to fuse with another gamete in order to form a new diploid cell, found in sexual eukaryotes. The number of chromosomes in a gamete is referred to as the 'haploid chromosome number'.

Gene – A DNA sequence carrying genetic information. A gene is located at a specific place – a locus – on a chromosome. Sometimes, when we talk about a 'gene', we assume that it contains relevant information that specifies some important biochemical/biological function. At other times, we are more general, and 'gene' stands just for genetic material of unknown significance at a well-specified chromosomal locus.

Genome – Used broadly, all the DNA that constitutes an individual's genotype. Used more specifically, the haploid genetic material that diploid individuals inherit two very similar copies of. Thus, humans inherit one genome from their mother via the egg and one genome from their father via the sperm. (Allotetraploid plants (see earlier) inherit four such genome copies, two via the egg cell of the mother plant and two via the fertilizing pollen.)

Genotype – The genetic composition of an individual organism. The term is often used with a restricted meaning to refer to the genetic composition at a specific locus or set of loci.

Haploid – An organism or a cell is said to be haploid if it contains only one copy of the species' genome.

Haplotype – The genetic composition of a unit (*e.g.* a gametic cell) with a haploid genome. Often, the word denotes the genetic composition of a particular chromosome stretch with many loci. Thus, a human inherits one haplotype from the mother and one haplotype from the father of every homologous chromosome. In the description of haplotypes, the order of the loci along the chromosome is often assumed known.

Hardy–Weinberg proportions (HW) – In a population where mating is at random and no disturbing factors occur, the probability that a homozygote for an allele with gamete frequency p is formed by fertilization is p^2. The probability that a heterozygote is formed with exactly one gene copy of the specified allelic type is $2p(1 - p)$. The probability that neither of the copies is of the specified allele is $(1 - p)^2$. (For more on this, see Background 2:1.) These probabilities constitute the Hardy–Weinberg proportions. In some genetics textbooks, a 'Hardy–Weinberg principle' or similar is discussed. By this is meant what we here call the Law of Constancy of Allele Frequencies. We prefer to treat this idea separately, so that no restriction of this law to diploidy – which is central to the arguments of Hardy and Weinberg – is implied.

Heritability – In this book, 'heritability' is used only in its narrow sense, to mean a measure (generally denoted h^2) of the proportion that the additive genetic variance constitutes of the total phenotypic variance. The heritability relates to a particular population in its specific environment and is not primarily a function of the investigated trait as such. For more on the meaning of 'additive genetic variance' and how it can be estimated, see any textbook on quantitative genetics.

Heterozygosity – See *Diversity*.

Inbreeding – A word with many related meanings. According to the usage that we adhere to here, inbreeding occurs when two related individuals produce offspring together. There is then a possibility that the offspring becomes homozygous by descent through inheriting identical copies of genetic material from its ancestor(s) via both its parents. In a very large, random-mating population, the possibility of inbreeding can be ignored, but since all real populations are finite and limited, all individuals are in a technical sense inbred. Normally, however, the term 'inbred' is only used in connection with matings between fairly close relatives (often ascertained via pedigree information). Inbreeding is generally measured by the inbreeding coefficient, F. A diploid individual's degree of inbreeding is the probability that it is homozygous by descent (according to known information) at a random locus in the genome. The degree to which a population is inbred is given by the mean of its members' inbreeding coefficients.

Indel – Genetic variation consisting of consecutive base pairs that have either been *ins*erted into or *del*eted from a DNA sequence. In a heterozygote for an indel, one allele has, thus, a sequence that is longer than that of the other.

Infinite sites model – An approximation of the structure of a gene and how this gene may change by mutation. Under this model, a gene is assumed to consist of an infinitely long sequence of sites. A new mutation will occur at a random position in this sequence and lead to a change in its state. Under these assumptions, the same mutation can never occur twice; every new mutation leads to a new allele for the gene and there may, in principle, be infinitely many alleles.

Law of Constancy of Allele Frequencies – The frequency of an allele in a population does not change in any systematic way when the genetic material is transmitted from one generation to the next, unless the allele is affected by some evolutionary force such as mutation, selection or segregation distortion. Given the existence of chance effects – genetic drift – during generational change, the law is formally better stated as saying that when no evolutionary forces exist and the Mendelian law of segregation holds (if relevant), the *expectation* of an allele's frequency is identical to its value in the previous generation. This genetic inertia constitutes the basis for the efficacy of the evolutionary forces, despite their sometimes small strength, in changing the composition of a population and its phenotypic properties.

Linkage – Two loci are linked if the recombination frequency between them is smaller than 0.5 (as applied to sexual species).

Linkage disequilibrium (LD) – The association (correlation) that may exist between alleles at different loci. The name is strange and unfortunate. Linkage disequilibrium can be measured in many ways, of which the most common is D (see Derivation 3). Two loci are said to be in linkage disequilibrium when one allele at the first and one at the second co-occur more (or less) often than expected by chance alone. Two loci that are not in linkage disequilibrium are said to be in linkage equilibrium. Linkage disequilibrium may occur between loci that are on different chromosomes, for example directly after a population has been formed by contributions from two distinct gene pools. The phenomenon is therefore not restricted to linked loci.

Load, genetic load – In a population where not all individuals have the maximally fit genotype, a proportion of the population – or a proportion of its reproductive capacity – is lost in every generation due to selection. This proportion is (by a somewhat strange convention) called the population's load. The loss may be due to deleterious mutant alleles causing genetic diseases – the mutation load. Or it may be due the reduced fitness of homozygotes when heterozygote advantage holds – the segregation load. In general, one expects that natural selection will decrease a population's load by secondary selection. Thus, for example, stronger and better repair systems have throughout life's existence evolved to reduce the frequency of deleterious mutations and thereby the mutation load.

Locus – A specific position along a chromosome. The physical extent of 'a locus' may vary with the context – sometimes it may refer to a single base pair, while at other times it may encompass a long string of genes (as when one talks about the HLA locus). Very often, the term refers to the chromosomal position of a particular functional gene.

Logistic equation, logistic growth – The logistic differential equation $\frac{dy}{dx} = b - ay$ (where $a, b > 0$) describes the process whereby a dependent variable, y, increases at a rate a for small x values, but with greater x values increases more slowly as the y variable approaches b/a; see Background 1:2. The logistic equation is commonly used to describe organism growth dependent upon some limiting resource. Sometimes it is preferable to use the equation in the form of a difference equation; for example, $N_{t+1} - N_t = R\left[1 - \frac{N_t}{K}\right]$, where $R, K > 0$.

Meiosis – The process by which sexual diploid organisms produce haploid cells (often in connection with gamete production). The halving of the genetic material in meiosis is a necessary balance to the fusing of cells and genetic material in fertilization.

Meiotic cycle – In sexually reproducing eukaryotes, the genetic material goes through a (in many way strange and unexpected) cycle. Fertilization leads to a diploid cellular state, which may be long or short.

Meiosis, almost always associated with recombination, then occurs, leading back to a haploid cell state. This haploid state may, again, be long or short, before two haploid cells fuse in a new fertilization event.

Mitochondrion, mainly used in plural form as *mitochondria* – A cellular organelle common to almost all eukaryotes. It contains its own specific DNA and is in animals and plants transmitted only via the egg cell (with, perhaps, some very rare examples of 'leakage' from the male gametic line).

Mitosis – The process by which the duplicated chromosomes in a cell separate in a highly ordered fashion during cell division. Mitosis ensures that the two daughter cells obtain exactly the same genetic material as the parental cell originally had.

Most recent common ancestor (MRCA) – Here applied to a sample of homologous sequences, and referring to the copy – somewhere back in time – from which all the copies in the sample are descended.

Neutrality, selective neutrality, effective neutrality – Neutrality holds when all genotypes have the same fitness. A particular genetic variation is effectively neutral if the population size is so small that the genetic drift overpowers the selective differences between the genotypes.

Phenotype – The outward appearance of an individual organism, nowadays extended to encompass physiological traits as well as behaviours (*i.e.* any property of the organism). An individual's genotype is one of the determinants of its phenotype.

Polymorphism, or more strictly *genetic polymorphism* – When two or more common variants occur at a locus. In practice, this is often interpreted to mean that the frequency of the most common type is less than 99% (sometimes a limit of 95% is used instead). A polymorphism is said to be protected if the relevant boundary equilibria are unstable. A polymorphism is balanced if it derives from contradictory evolutionary forces that cause the population to move to a stable internal equilibrium.

Population subdivision – A population is said to be subdivided if it consists of a group of more or less reproductively separated subpopulations. The degree of genetic differentiation between two or many subpopulations is commonly measured by the index F_{ST} – see Derivation 7.

Prokaryote – See *Eukaryote.*

Quantitative genetics – The part of genetics which studies the inheritance of traits based on the action of many, not directly specified, genes.

Random mating – When males and females mate independent of origin and genotype. Often, the term is used when 'random union of gametes' would be biologically more meaningful. Random mating is always defined with respect to a particular mating population. Thus, in a reproductively subdivided population, random mating may very well take place *within* the subpopulations but not *between* them. See Derivation 7.

Recessive – See *Dominance.*

Recombination – We apply the term 'recombination' in this book to what happens during the meiotic process in organisms that regularly go through the meiotic cycle. Recombination occurs between two loci, A and B, if individuals created from the fertilization of A_1B_1 gametes by A_2B_2 gametes (the parental types) produce these gametes but also A_1B_2 and A_2B_1 gametes (the recombinant types). Recombination between different genetic materials may also occur in other kinds of organisms (*e.g.* bacteria), but it takes then a less well-regulated form.

Recombination frequency – The proportion of recombinant-type gametes produced by meiosis in an individual that is heterozygous for two investigated loci (see *Recombination*).

Recursion system – An equation (or a set of equations) that describes how a system at a given moment in time depends on the state of the system at some earlier time. Here, we only use recursion systems where the composition of a population is a function of the immediately preceding generation. Such processes are said to exhibit the Markov property.

Segregation distortion – The most important of Mendel's rules states that a heterozygote produces the two possible kinds of gametes with equal frequencies. When this does not hold, segregation is said to be distorted. One cause of segregation distortion may be that one chromosome in a homologous pair 'pushes for itself' during meiosis: so-called 'meiotic drive'. In a classic example of segregation distortion, male mice with genetic constitution $+/t$ produce sperm with almost only the t allele.

Sex chromosome – A chromosome inherited in a sex-specific manner and directly or indirectly involved in the determination of sex.

Sex ratio – The ratio between the sexes in a bi-sexual population. In animals, the sex ratio is normally given as the proportion of males divided by the proportion of females.

Single nucleotide polymorphism (*SNP*, generally pronounced 'snip') – Strictly defined, there is a SNP at a specific site in a DNA sequence in a given population if the frequency of the most common nucleotide at this site is less than 99%. Often, the term is used more loosely: when two homologous sequences are compared, any observed difference between them is called a SNP (without anything being known about the population frequencies). In practice, it is rare to find more than two nucleotide alternatives at a particular site when DNA sequences are compared, unless they separated *very* far back in history. SNPs are therefore often implicitly assumed to be bivariate.

Stability – Used of equilibria to dynamic processes. An equilibrium point is globally stable if the system will move towards this point from any valid starting position. The equilibrium point is locally stable if the system will move towards this point from a starting position slightly perturbed away from the equilibrium point. All globally stable equilibria are also locally stable.

Virus – Viruses are not cellular organisms, and it is a matter of definition whether they are regarded as living matter or not. They contain, however, DNA (or RNA, which as genetic material is very similar), and their populations can often be analysed with standard methods used in population genetics.

Wahlund effect, Wahlund's variance – In a reproductively subdivided population, the total population will never show as many heterozygotes as expected from Hardy–Weinberg considerations, given the overall frequencies of the alleles. The size of this so-called Wahlund effect depends on the variance in the allele frequencies over the different subpopulations, called Wahlund's variance. See Derivation 7.

Wright–Fisher population – An idealized population in which reproduction is assumed to occur in a very simple way, making the transmission of genetic material between generations easily amenable to mathematical analysis – see Derivation 4.

ANSWERS

Derivation 1

1. For the first set of mutation rates, $p_{eq} = 3 \cdot 10^{-6}/(1 \cdot 10^{-6} + 3 \cdot 10^{-6}) = 3/4 = 0.75$. For the second set, $p_{eq} = 3 \cdot 10^{-8}/(1 \cdot 10^{-6} + 3 \cdot 10^{-8}) = 3/103 \approx 0.029$.

2. Note that $p = 0.625$ is the exact midpoint between p_{eq} and the start at $p_0 = 0.50$. Thus, according to (1.6), the time to get there is equal to $\ln(2)/(3 \cdot 10^{-6} + 10^{-6}) = \ln(2)/4 \cdot 10^{-6} \approx 0.173 \cdot 10^6 = 1.73 \cdot 10^5$ generations.

3. Since, according to expression (1.5b), $p_t = p_{eq} + (p_0 - p_{eq})e^{-(\mu_{12}+\mu_{21})t}$, we have that $p_t - p_{eq} = (p_0 - p_{eq})e^{-(\mu_{12}+\mu_{21})t}$ and that $(p_t - p_{eq})/(p_0 - p_{eq}) = e^{-(\mu_{12}+\mu_{21})t}$. From this, it follows that $\ln[(p_t - p_{eq})/(p_0 - p_{eq})] = -(\mu_{12} + \mu_{21})t$, which gives the required formula $t = -\ln[(p_t - p_{eq})/(p_0 - p_{eq})]/(\mu_{12} + \mu_{21})$. According to this formula, it takes $-\ln[(0.7 - 0.75)/(0.5 - 0.75)]/(10^{-6} + 3 \cdot 10^{-6}) = -\ln[-0.05/-0.25]/4 \cdot 10^{-6} = -\ln(0.20)/4 \cdot 10^{-6} = \ln(5)/4 \cdot 10^{-6} \approx 0.402 \cdot 10^6 = 4.02 \cdot 10^5$ generations to go from $p_0 = 0.50$ to $p_t = 0.70$.

4. Expression (1.1) changes to $p_{t+1} = p_t(1 - \mu_{12})$. By substituting $p_{t+1} = p_t = p_{eq}$ in this equation, it can be seen that $p_{eq} = 0$ is the only possible equilibrium to the system. This equilibrium point, which corresponds to the loss of allele A_1 from the population, is obviously stable, since it will be reached from all possible starting frequencies of p. The process towards the loss of A_1 is described by noting that $p_t = p_{t-1}(1 - \mu_{12}) = p_{t-2}(1 - \mu_{12}) \cdot (1 - \mu_{12}) = (1 - \mu_{12})^2 p_{t-2} = (1 - \mu_{12})^t p_0$, which shows that the allele frequency p will constantly decrease over time, with 0 as its lower limit. (With respect

Understanding Population Genetics, First Edition. Torbjörn Säll and Bengt O. Bengtsson.
© 2017 John Wiley & Sons Ltd. Published 2017 by John Wiley & Sons Ltd.

to the deletions used as an example, this argument assumes, of course, that there is no selection counteracting the mutations.)

5. From the assumptions summarized in Figure 1.2, it follows that
$$p(A)_t = (1 - \alpha - 2\beta)p(A)_{t-1} + \alpha p(G)_{t-1} + \beta p(T)_{t-1} + \beta p(C)_{t-1}.$$
From the symmetry in the mutation scheme, it is reasonable to guess that the frequencies of the four bases will evolve towards $p(A) = p(G) = p(C) = p(T) = 1/4$. That this is indeed an equilibrium point to the system is supported by the fact that if in the expression just derived for $p(A)_t$ we insert in the right-hand side that $p(A)_{t-1} = p(G)_{t-1} = p(C)_{t-1} = p(T)_{t-1} = 1/4$, then we find that $p(A)_t = (1 - \alpha - 2\beta)/4 + \alpha/4 + \beta/4 + \beta/4 = 1/4$. It is thus clear that with this set of frequencies for the four bases, the frequency of A, at least, will not change over time. That the same holds also for the other nucleotides is trivial to show.

Derivation 2

1. The mutation rates in Figure 1.1 are very small, while the selective forces in Figure 2.1 are quite strong. The rate of change in the allele frequency is therefore much smaller in the first case than in the second. While mutation rates are almost always small, selection coefficients may range from substantial (*e.g.* during ecological catastrophes or with respect to severe genetic diseases) to imperceptibly minute (*e.g.* between different nucleotide triplets coding for the same amino acid). For the second part of the question, we note that mutation has the ability to create new variation. Thus, even if one starts with a genetically monomorphic population in the first case, the mutations assumed in the process will take the population away from this state with no variation. Selection, on the other hand, does not have this ability. It only changes the frequencies of existing genotypes and alleles. In a population with no variation, selection – in itself – will not change its monomorphic genetic composition.

2. Using the fact that $q = 1 - p$, one gets that $W = 1 - sp^2 - t(1 - p)^2$
$$= 1 - sp^2 - t(1 - 2p + p^2) = 1 - sp^2 - t + 2tp - tp^2 = 1 - t + 2tp$$
$- (s + t)p^2$. The derivative of W with respect to p, $\frac{dW}{dP}$, is equal to $2t - (s + t) \cdot 2p$. Obviously, this derivative equals 0 only when

$p = t/(s + t)$. The derivative is positive at $p = 0$ and is a decreasing function of p; thus, $p = t/(s + t)$ must represent a maximum to the W function. For increasing values on p, the population fitness will therefore increase from $1 - t$ at $p = 0$ to its maximum at $p = t/(s + t)$, and then decrease towards $1 - s$ for $p = 1$. The internal equilibrium under heterozygote advantage is obviously the point where this random-mating population has its highest possible mean fitness. (If the population were to become instantaneously asexual, then the homozygotes would disappear and the frequency of heterozygotes would increase until the heterozygotes became the only genotype in the population; the mean population fitness would then be exactly 1 and there would be no segregation load.)

3. Under the first fitness regime, we have that $W = p^2 + 2pq(1 - s) + q^2(1 - 2s) = 1 - 2sq(p + q) = 1 - 2sq$ and that $p' = [p^2 + 1/2 \cdot 2pq \cdot (1 - s)]/W = p[p + q(1 - s)]/W = p(1 - sq)/(1 - 2sq)$. Under the second regime, $W = p^2(1 + 2s) + 2pq(1 + s) + q^2 = 1 + 2sp(p + q) = 1 + 2sp$ and $p' = [p^2 \cdot (1 + 2s) + 1/2 \cdot 2pq \cdot (1 + s)]/W = p[(1 + 2s)p + (1 + s)q]/W = p(1 + 2sp + sq)/(1 + 2sp)$. If s is a small value, we can ignore terms of size hs^2 and approximate $1/(1+as)$ with $1 - as$ (as explained by result (2) in Background 2:2). For the first case, this implies that $p' \approx p(1 - sq)(1 + 2sq) \approx p(1 - sq + 2sq) = p(1 + sq) = p[1 + s(1 - p)]$. For the second, $p' \approx p(1 + 2sp + sq)(1 - 2sp) \approx p(1 + 2sp + sq - 2sp) = p(1 + sq) = p[1 + s(1 - p)]$. The two fitness schemes lead, thus, to the same dynamics for the positive mutation as long as the fitness advantage s is small.

4. The formula used in the main derivation says that $p' = p(1 - sp)/[1 - sp^2 - t(1 - p)^2]$. In this formula, we shall substitute $-s/(2 - s)$ for s and $s/(2 - s)$ for t. This gives $p' = p\left(1 + \frac{sp}{2-s}\right) \bigg/ \left[1 + \frac{sp^2}{2-s} - \frac{s(1-p)^2}{2-s}\right] = \frac{p(2-s+sp)}{2-s+sp^2-s(1-p)^2}$. Now we do some rewriting, using that $q = 1 - p$: the enumerator can be written $p[2 - s(1 - p)] = p(2 - sq)$, while the denominator can be written $2 - s + sp^2 - s(1 - 2p + p^2) = 2 - s + sp^2 - s + 2sp - sp^2 = 2 - 2s + 2sp = 2[1 - s(1 - p)] = 2(1 - sq)$. Thus, we have that $p' = \frac{p(2-sq)}{2(1-sq)} = \frac{p(1-sq/2)}{1-sq}$, just as required.

5. According to its definition, $Cov[X, Y] = E[XY] - E[X] \cdot E[Y]$ (see Background 7:1). We start by finding expressions for the

three factors to the right of the equality sign. If X is the number of copies of the A_1 allele that an individual carries in the assumed random-mating population with heterozygote advantage, then $E[X] = p^2 \cdot 2 + 2pq \cdot 1 + q^2 \cdot 0 = 2p(p + q) = 2p$, where q stands for $1 - p$. Similarly, if Y is the fitness of the individuals in the population, then $E[Y] = p^2 \cdot (1 - s) + 2pq \cdot 1 + q^2 \cdot (1 - t) = (p^2 + 2pq + q^2) - sp^2 - tq^2 = (p + q)^2 - sp^2 - tq^2 = 1 - sp^2 - tq^2$. Finally, the expectation of the product of these values is $E[XY] = p^2 \cdot 2 \cdot (1 - s) + 2pq \cdot 1 \cdot 1 + q^2 \cdot 0 \cdot (1 - t) = 2(1 - s)p^2 + 2pq$. With these inputs, the requested covariance becomes $Cov[X, Y] = E[XY] - E[X]E[Y] = 2(1 - s)p^2 + 2pq - 2p \cdot (1 - sp^2 - tq^2) = 2p(p - sp + q - 1 + sp^2 + tq^2) = 2p(-sp + sp^2 + tq^2) = 2p[-sp(1 - p) + tq^2] = 2pq(-sp + tq)$. From expression (2.1), we have that $p' = p(1 - sp)/[1 - sp^2 - tq^2]$. By subtracting p from both sides, we get $p' - p = \Delta p = \frac{p(1-sp)}{1-sp^2-tq^2} - p = \frac{p(1-sp)-p+sp^3+tpq^2}{1-sp^2-tq^2} = p\frac{1-sp-1+sp^2+tq^2}{1-sp^2-tq^2} = p\frac{-sp(1-p)+tq^2}{1-sp^2-tq^2} = pq\frac{-sp+tq}{1-sp^2-tq^2} = \frac{2pq(-sp+tq)}{2(1-sp^2-tq^2)}$.

Thus, it follows that $\Delta p = \frac{Cov[X,Y]}{2W}$ and – since W always is positive – that Δp equals 0 when $Cov[X, Y]$ equals 0, and *vice versa*.

Derivation 3

1. In the present generation, the linkage disequilibrium, which we denote D_0, is equal to $0.62 \cdot 0.13 - 0.09 \cdot 0.16 = 0.0662$. In this generation, we also have the following allele frequencies: $p_0(A_1) = 0.62 + 0.09 = 0.71$, $p_0(A_2) = 0.13 + 0.16 = 0.29$, $p_0(B_1) = 0.62 + 0.16 = 0.78$ and $p_0(B_2) = 0.09 + 0.13 = 0.22$. In the next generation, according to the gamete production table, the haplotype frequencies will be:
$P'(A_1B_1) = 0.62^2 + 2 \cdot 0.62 \cdot 0.13 \cdot 0.85/2 + 2 \cdot 0.09 \cdot 0.16 \cdot 0.15/2 + 2 \cdot 0.62 \cdot (0.09 + 0.16)/2 \approx 0.6101$
$P'(A_1B_2) = 0.09^2 + 2 \cdot 0.09 \cdot 0.16 \cdot 0.85/2 + 2 \cdot 0.62 \cdot 0.13 \cdot 0.15/2 + 2 \cdot 0.09 \cdot (0.62 + 0.13)/2 \approx 0.0999$
$P'(A_2B_1) = 0.16^2 + 2 \cdot 0.09 \cdot 0.16 \cdot 0.85/2 + 2 \cdot 0.62 \cdot 0.13 \cdot 0.15/2 + 2 \cdot 0.16 \cdot (0.62 + 0.13)/2 \approx 0.1699$

$P'(A_2B_2) = 0.13^2 + 2 \cdot 0.62 \cdot 0.13 \cdot 0.85/2 + 2 \cdot 0.09 \cdot 0.16 \cdot$
$0.15/2 + 2 \cdot 0.13 \cdot (0.09 + 0.16)/2 \approx 0.1201$

The linkage disequilibrium in the next generation will, thus, be $D_1 = 0.6101 \cdot 0.1201 - 0.0999 \cdot 0.1699 \approx 0.0563$. This conforms with the expected value according to (3.9a), which is $(1 - 0.15) \cdot 0.0662 \approx 0.0563$. The allele frequencies in this new generation, calculated from the haplotype frequencies, will be $p_1(A_1) = 0.6101 + 0.0999 = 0.7100 = 0.71$, $p_1(A_2) = 0.1699 + 0.1201 = 0.2900 = 0.29$, $p_1(B_1) = 0.6101 + 0.1699 = 0.7800 = 0.78$ and $p_1(B_2) = 0.0999 + 0.1201 = 0.2200 = 0.22$.

These values also conform to what we expect, namely that recombination *per se* does not change any allele frequencies in the absence of other evolutionary forces.

2. In the starting generation, the frequencies of the two kinds of double heterozygotes are $2 \cdot P(A_1B_1) \cdot P(A_2B_2) = 2 \cdot 0.62 \cdot 0.13 = 0.1612$ and $2 \cdot P(A_1B_2) \cdot P(A_1B_2) = 2 \cdot 0.09 \cdot 0.16 = 0.0288$. The difference between these two values is 0.1324.

In the following generation, the frequencies of the two kinds of double heterozygotes are $2 \cdot P(A_1B_1) \cdot P(A_2B_2) - 2 \cdot 0.6101$ $0.1201 \approx 0.1465$ and $2 \cdot P(A_1B_2) \cdot P(A_1B_2) = 2 \cdot 0.0999 \cdot 0.1699 \approx 0.0339$; their difference is now 0.1126.

It seems reasonable to believe that the difference between these two frequencies will decrease to 0 with time. That this indeed is the case can be shown by expressing the frequencies of the two kinds of heterozygotes in any population in general terms with the help of D. Thus, we have that the frequency of genotype A_1B_1/A_2B_2 equals $2 \cdot P(A_1B_1) \cdot P(A_2B_2) = 2 \cdot [p(A_1) \cdot p(B_1) + D] \cdot [p(A_2) \cdot p(B_2) + D]$, while the frequency of genotype A_1B_2/A_2B_1 equals $2 \cdot P(A_1B_2) \cdot P(A_2B_1) = 2 \cdot [p(A_1) \cdot p(B_2) - D] \cdot [p(A_2) \cdot p(B_1) - D]$. The difference between these two frequencies equals $2 \cdot [p(A_1) \cdot p(B_1) + D] \cdot [p(A_2) \cdot p(B_2) + D] - 2 \cdot [p(A_1) \cdot p(B_2) - D] \cdot [p(A_2) \cdot p(B_1) - D] = 2 \cdot p(A_1) \cdot p(A_2) \cdot p(B_1) \cdot p(B_2) + 2 \cdot D \cdot p(A_1) \cdot p(B_1) + 2 \cdot D \cdot p(A_2) \cdot p(B_2) + D^2 - (-2) \cdot p(A_1) \cdot p(A_2) \cdot p(B_1) \cdot p(B_2) + 2 \cdot D \cdot p(A_1) \cdot p(B_2) + 2 \cdot D \cdot p(A_2) \cdot p(B_1) - D^2 = 2D[p(A_1) \cdot p(B_1) + p(A_2) \cdot p(B_2) + p(A_1) \cdot p(B_2) + p(A_2) \cdot p(B_1)] = 2D\{p(A_1)[p(B_1) + p(B_2)] + p(A_2)[p(B_2) + p(B_1)]\} = 2D[p(A_1) + p(A_2)] = 2D.$

Since the linkage disequilibrium, D, decreases towards 0 with time, so too must the difference in frequency between the two types of double heterozygotes. (That this difference in any given generation equals twice the linkage disequilibrium is illustrated by the numerical results we have just obtained: in the starting generation, $D = 0.0662$, while the difference between the two genotypes frequencies is 0.1324.)

3. A quick and intuitive (and not necessarily wrong) solution is to start by finding the largest possible values for $P(A_1B_1)$ and $P(A_2B_2)$. With the given numerical assumptions, these are obviously 0.6 for $P(A_1B_1)$ and 0.3 for $P(A_2B_2)$. It then follows that $P(A_1B_2)$ must be 0 and that $P(A_2B_1)$ must be 0.1. With these haplotype frequencies, D_{max} becomes $0.6 \cdot 0.3 - 0 \cdot 0.1 = 0.18$.

A formally more satisfying solution is to consider a value x such that $p(A_1B_1) = 0.6 - x$. It then follows that $P(A_2B_1) = 0.7 - (0.6 - x) = 0.1 + x$, $P(A_1B_2) = x$, and that $P(A_2B_2) = 0.3 - x$ (the value x must thus belong to the interval $0 \leq x \leq 0.3$). With these haplotype frequencies, the linkage disequilibrium D becomes $(0.6 - x)(0.3 - x) - (0.1 + x)x = 0.18 - (0.6 + 0.3)x + x^2 - 0.1 \cdot x - x^2 = 0.18 - x$. Given the constraints on x, this value is numerically maximized at $x = 0$, making D_{max} equal to 0.18.

4. D_{max} is defined by the allele frequencies in the population. Since these — given selective neutrality — are not assumed to change over time, D_{max} will always stay the same. According to its definition, D' equals D/D_{max}, while according to our main result (see expression (3.9)), D_{t+1} equals $(1 - r)D_t$; it therefore follows that $D_{t+1}/D_{max} = (1 - r)D_t/D_{max}$, or, in other words, that $D'_{t+1} = (1 - r)D'_t$. D' is, thus, expected to decrease towards 0 at the same rate as D, as long as the process is affected by recombination alone. Note that the prime sign in D' has nothing to do with the meaning 'this parameter in the next generation', which it has in other contexts.

5. With obvious notations we have that initially in the hybrid population $P(A_1B_1) = 0.6$ and $P(A_2B_2) = 0.4$, while $P(A_1B_2) = P(A_2B_1) = 0$. This makes the initial linkage disequilibrium, D_0, equal to $0.6 \cdot 0.4 - 0 \cdot 0 = 0.24$. In the question we are told that the current linkage

disequilibrium, which we designate D_t, equals 0.09. This change in D from 0.24 to 0.09 from generation 0 to generation t should follow the dynamics given by expression (3.9b), *i.e.* that $D_t = (1 - r)^t D_0$; in this case the formula implies that $0.09 = (1 - 0.08)^t \cdot 0.24$. By rearrangement we get that $0.92^t = 0.09/0.24 = 0.375$, and that therefore $t = \ln(0.375)/\ln(0.92) \approx 11.76$. Thus, the hybrid population should have been founded about twelve generation ago.

Derivation 4

1. (a) There are 4000 autosomal gene copies per generation (2N). Thus, according to expression (4.1), the mean time in generations to coalescence for two randomly drawn gene copies is 4000. (b) With a balanced sex ratio, there are 1000 Y chromosomes in the population. Their inheritance among males follows the straightforward Wright–Fisher transmission logic for haploid genetic material. Thus, the time to coalescence is 1000 generations. (c) Males are dead ends for mitochondria, so in the transmission history of the mitochondria, only the 1000 females per generation matter. A first answer to the question is therefore 1000 generations, in correspondence with the preceding answer. This is, however, not quite correct, since there are many copies of the mitochondrial chromosome in every egg cell. The mitochondrial DNA in an individual is therefore not necessarily homogenous; individuals may be what is called 'heteroplastic'. If this is taken into account, the mean time to coalescence increases – by how much depends on the segregation and homogenization of the mitochondrial DNA during cell divisions in development.

2. According to the formula given in the discussion, $N_e = \frac{4 N_m N_f}{N_m + N_f}$, so we get that $N_e = \frac{4 \cdot 10 \cdot 400}{10 + 400} = \frac{16000}{410} \approx 39.0$, *i.e.* approximately 40. Thus: (a) about 80 generations; (b) about 10 generations; (c) about 400 generations (though with the same caveat about heteroplasmy as discussed in Answer 1). The answer to (a) is based on considerations for a diploid Wright–Fisher population, while the answers to (b) and (c) are based on the formula for a haploid Wright–Fisher population.

3. The formula given in the discussion for the population size of a regularly fluctuating diploid population can obviously be used. According

to this formula, $\frac{1}{N_e} = \frac{1}{2}\left[\frac{1}{N_s} + \frac{1}{N_l}\right] = \frac{1}{2}\left[\frac{1}{10^4} + \frac{1}{10^8}\right] = \frac{1}{2}\left[\frac{10000+1}{10^8}\right] = \frac{10001}{2 \cdot 10^8}$, which leads to $N_e \approx 19998$. The effective size of the population is, thus, about 20 000, or $2 \cdot 10^4$. Thus, if there is almost no genetic drift in every second generation, then the effective population size becomes close to twice the smaller experienced population size.

4. The mean time to coalescence must obviously be smaller than $2N$. This is easily seen by revisiting the principles used in the chapter's main derivation. If we follow two gene copies back in time, then the probability that they will coalesce in the preceding generation becomes greater and greater with a smaller and smaller size of the anterior populations. This leads to a shorter mean time to coalescence compared to if the population size were constant. Since shorter times to coalescence mean – on average – less time for mutations to happen, this result makes us expect that two randomly drawn gene copies from a constant population will be more genetically variable than two gene copies drawn from an expanding population that is currently of the same size. More on this in the Derivation 5.

5. (a) To calculate the probabilities asked for, we first follow one of the gene copies back to the preceding generation. Then, the probability that the other two gene copies are copies of exactly the same preceding copy becomes $P(2\ coalescence\ events) = \frac{1}{2N} \cdot \frac{1}{2N} = \frac{1}{4N^2}$. The probability that the first and the second copy do not coalesce, and neither does the third copy with either of the other two, becomes $P(0\ coalescence\ events) = \frac{2N-1}{2N} \cdot \frac{2N-2}{2N} = \left[1 - \frac{1}{2N}\right]\left[1 - \frac{2}{2N}\right] = 1 - \frac{3}{2N} + \frac{2}{4N^2}$. There are three ways by which two of the gene copies may coalesce but not the third, which gives the associated probability $P(1\ coalescence\ event) = 3 \cdot \frac{1}{2N} \cdot \left[1 - \frac{1}{2N}\right] = \frac{3}{2N} - \frac{3}{4N^2}$.

(It is always important to check, and – yes, indeed – these three values add to 1, as they should.)

(b) When N is so large that we can ignore $\frac{1}{N^2}$, the three probabilities simplify to $P(0\ coalescence\ events) \approx 1 - \frac{3}{2N}$, $P(1\ coalescence\ events) \approx \frac{3}{2N}$ and $P(2\ coalescence\ events) \approx 0$.

These results fit exactly the more general derivation in the text where the behaviour of n gene copies is analysed.

Derivation 5

1. (a) According to expression (5.2b), the genetic variation in a stable diploid population of size N is $H \approx \frac{4N\mu}{1+4N\mu}$. This implies, in the outlined example, that $H_0 \approx \frac{4 \cdot 10^7 \cdot 0.14 \cdot 10^{-6}}{1+4 \cdot 10^7 \cdot 0.14 \cdot 10^{-6}} = \frac{5.6}{1+5.6} = \frac{5.6}{6.6} \approx$ 0.848, where H_0 is the genetic variation in the population before the collapse.

   (b) The genetic variation in the first generation after the collapse is $H_1 \approx \left[1 - \frac{1}{20000}\right] \cdot H_0 = \left[1 - \frac{1}{20000}\right] \cdot \left[\frac{5.6}{6.6}\right]$, which is extremely close to $\frac{5.6}{6.6} \approx 0.848$.

   (c) The equilibrium level of the genetic variation for a population of diploid size $N = 10\ 000$ is $H_{eq} = \frac{4 \cdot 10^4 \cdot 0.14 \cdot 10^{-6}}{1+4 \cdot 10^4 \cdot 0.14 \cdot 10^{-6}} = \frac{0.056}{1+0.056} = \frac{0.056}{1.056} \approx 0.053$. According to expression (5.5b), $H_t \approx H_{eq} + (H_0 - H_{eq})e^{-\left(2\mu + \frac{1}{2N}\right)t}$, and we therefore have that t generations after the environmental collapse, the genetic variation should be: $H_{eq} + (H_0 - H_{eq})e^{-\left(2\mu + \frac{1}{2N}\right)t} = \frac{0.056}{1.056} + \left(\frac{5.6}{6.6} - \frac{0.056}{1.056}\right)e^{-\left(2 \cdot 0.14 \cdot 10^{-6} + \frac{1}{2 \cdot 10000}\right)t} \approx 0.053 + (0.848 - 0.053)$ $e^{-50.28 \cdot 10^{-6}t} = 0.053 + 0.795e^{-50.28 \cdot 10^{-6}t}$.

   Thus, $H_{100} = 0.053 + 0.795e^{-0.005028} \approx 0.844$, while $H_{1000} = 0.053 + 0.795e^{-0.05028} \approx 0.809$ and $H_{1000000} = 0.053 + 0.795e^{-50.28} \approx 0.053$.

2. The long-term equilibrium genetic variation is now $H_{eq} = \frac{4N\mu}{1+4N\mu} = \frac{4 \cdot 10^6 \cdot 0.14 \cdot 10^{-6}}{1+4 \cdot 10^6 \cdot 0.14 \cdot 10^{-6}} = \frac{0.56}{1+0.56} = \frac{0.56}{1.56} \approx 0.359$. The level of variation t generations after the population has reached its stable size is $H_{eq} + (H_0 - H_{eq})e^{-\left(2\mu + \frac{1}{2N}\right)t} = \frac{0.56}{1.56} + \left(0.001 - \frac{0.56}{1.56}\right)e^{-\left(2 \cdot 0.14 \cdot 10^{-6} + \frac{1}{2 \cdot 10^6}\right)t} \approx 0.359$ $+ (0.001 - 0.359)e^{-0.78 \cdot 10^{-6}t} = 0.359 - 0.358e^{-0.78 \cdot 10^{-6}t}$.

   Thus, $H_{100} = 0.359 - 0.358e^{-0.000078} \approx 0.001 = 0.1\%$, $H_{1000} = 0.359 - 0.358e^{-0.00078} \approx 0.001 = 0.1\%$ (*i.e.* almost the same), while $H_{1000000} \approx 0.359 - 0.358e^{-0.78} \approx 0.195 = 19.5\%$. This and the preceding question illustrate how slowly the genetic diversity of a population is expected to change when the mutation rate is small and the population size is substantial.

3. Let the seeds from the head constitute what we call 'the new population'. If we take a gene copy from this new population, then the probability is 1/2 that it derives from the mother plant and 1/2 that it derives from one of the father plants that provided the fertilizing pollen. If we now pick a second gene copy from the new population, then the probability that it comes from the same father plant as the first gene copy is so small that it can be ignored, given the assumption about the population being very large. That this second copy comes from the mother plant is still 1/2. Assume that the genetic variation in the original population is H_0. If zero or one of the picked gene copies is derived from the mother, then their probability of being different is obviously H_0. If both of them come from the mother – with probability 1/4 – then the chance that they are genetically identical is 1/2; otherwise, their genetic diversity is H_0, as it is for all random gene copies in this generation. Thus, the probability that the two randomly picked gene copies from the new generation are different is $\left(1 - \frac{1}{4}\right) H_0 + \frac{1}{4}\left(\frac{1}{2} \cdot 0 + \frac{1}{2}H_0\right) = \frac{7}{8}H_0$. This answer shows that a remarkably large part of the genetic variation in a random-mating population is retained among the offspring of an outbreeding mother.

4. According to its definition, $H_T = 1 - p^2 - q^2 = 2pq$. Sampling one individual from the population may lead to three different outcomes: the individual may be $A_1 A_1$, $A_1 A_2$ or $A_2 A_2$, and the probabilities of these outcomes are p^2, $2pq$ and q^2, respectively. The H values in these three cases are obviously 0, $1/2$ and 0. Thus, $E[H_S] = 0 \cdot p^2 + 1/2 \cdot 2pq + 0 \cdot q^2 = pq = H_T/2$.

5. The genetic variation is determined by the interplay between mutation in the studied DNA sequence and the genetic drift that occurs in the stable population of (effectively) 50 000 chloroplasts. The probability of no mutation in the sequence during one generation is $(1 - 3 \cdot 10^{-9})^{10000} \approx 1 - 10000 \cdot 3 \cdot 10^{-9} = 1 - 3 \cdot 10^{-5}$, so the mutation frequency for the sequence can be taken to be $3 \cdot 10^{-5}$. (The assumption that every mutation should be of a new

type is obviously fulfilled to a high degree of approximation.) According to the formula for the equilibrium level of genetic diversity between mutation and drift in haploids, we therefore have that $H_{eq} = \frac{2 \cdot 50000 \cdot 3 \cdot 10^{-5}}{1 + 2 \cdot 50000 \cdot 3 \cdot 10^{-5}} = \frac{3}{4} = 0.75$. The haploid version of expression (5.5b) says that $H_t = H_{eq} + (H_0 - H_{eq})e^{-\left(2\mu + \frac{1}{N}\right)t}$, and now we want to know the number of generations, t, before the level of variation is $H_t = 0.05$, given that $H_0 = 0$ and $H_{eq} = \frac{3}{4}$.

Since, with these assumptions, $0.05 = \frac{3}{4} + \left(0 - \frac{3}{4}\right)e^{-\left(2\mu + \frac{1}{N}\right)t}$, it follows that $e^{-\left(2\mu + \frac{1}{N}\right)t} = \frac{14}{15}$ and $t\left(2\mu + \frac{1}{N}\right) = \ln 15 - \ln 14$. Thus, $t = \frac{\ln 15 - \ln 14}{2 \cdot 3 \cdot 10^{-5} + (1/50000)} = \frac{(\ln 15 - \ln 14) \cdot 10^5}{6 + 2} \approx 862$.

Derivation 6

1. We take the statement that the population size is large to imply that $2Ns > 2$. The probability of fixation of the new positive mutation is therefore, according to approximation (6.11b), equal to the selective advantage, s. For a neutral mutation to have the same probability of fixation, its initial frequency should therefore be s (i.e. half a percent).

2. Look at the relevant formula (6.9) for the probability of fixation of an allele with frequency x and selective advantage s! In the two described cases, the numerator – determined by $2Nsx$ – remains the same. The denominator, however, is in the first case equal to $1 - e^{-0.02 \cdot N}$ and in the second equal to $1 - e^{-0.01 \cdot N}$. The second value is always smaller than the first. Thus, it is more likely that a mutation with $x = 0.02$ and $s = 0.005$ goes to fixation than a mutation with $x = 0.01$ and $s = 0.01$.

3. To be on the safe side, and to use the same formula for all cases, we do not approximate, but utilize expression (6:12a) throughout: $f\left(\frac{1}{2N}\right) = \frac{e^{|s|} - 1}{e^{2N|s|} - 1}$. We then find that the fixation probability of a new deleterious mutation is equal to $0.45 \cdot 10^{-4}$ for $s = -10^{-5}$, is equal to $0.16 \cdot 10^{-4}$ for $s = -10^{-4}$ (only slightly smaller) and is equal to $0.21 \cdot 10^{-11}$ for $s = -10^{-3}$ (*very* much smaller). The collapse in the fixation probability when $N|s|$ becomes appreciably greater than 1 is clearly illustrated.

4. For fixation to happen, a positive mutation must occur in the population that is not lost by chance. According to expression (6.11b), the probability of fixation of a new mutation with the specified advantage is 0.01. This implies that about 100 mutations must occur in the population before the one comes about that will go to fixation. (Here, we have yet another example of the geometric distribution.) Since the number of gene copies in the population is rather low, 10 000, and the mutation rate is small, 10^{-7}, new single mutations will be found in the population with an interval of about 1000 generations. Thus, the relevant mutation occurs once per 1000 generations. The population must therefore 'wait' for an average of $100 \cdot 1000 = 100\,000$ generations before the crucial mutation event leading to fixation occurs. The time for this mutation to actually go to fixation will not be particularly long. The outlined situation is therefore mainly what in probability theory is called a waiting-time problem.

5. Look at the proof leading up to the function for the probability of fixation for an allele with frequency x, $f(x)$. It was constructed using results concerning selection, summarized by $E[\delta] = sx(1-x)/2$, and drift, summarized by $E[\delta^2] = x(1-x)/2N$. Even if our interest now is in the probability of loss, these expressions summarizing the dynamics of the evolutionary process remain exactly the same. Thus, the function that we are looking for, $l(x)$, is found by solving the same differential equation as before – expression (6.1) – with the sole change that the name of the function, $f(x)$, is altered to $l(x)$. Thus, $2Ns\frac{dl(x)}{dx} + \frac{d^2l(x)}{dx^2} = 0$. At first, it may seem strange that the same differential equation can be used to find $f(x)$ and $l(x)$, but remember that solving this equation leads to a set of functions! To determine the relevant solution among all the alternatives in this set, the information about the boundary conditions should be used. And the boundary conditions are this time completely different, since we obviously now have that $l(0) = 1$ and that $l(1) = 0$, saying that a lost allele is certainly lost and a fixed allele certainly cannot become lost. It is thus the boundary conditions and not the differential equation that differ between the cases of fixation versus loss.

A check of what the new boundary conditions say gives us the following: If, in the full set of solution described by expression

[244]

(6.8), we use the first boundary condition, then we have that $-\frac{C}{2Ns} + C^* = 1$, which implies that $C^* = 1 + \frac{C}{2Ns}$. The second boundary condition gives that $C\left(-\frac{1}{2Ns}\right)e^{-2Ns} + C^* = 0$, which implies that $C^* = \frac{C}{2Ns}e^{-2Ns}$. Taking these results together implies that $C = -\frac{2Ns}{1-e^{-2Ns}}$ and $C^* = -\frac{e^{-2Ns}}{1-e^{-2Ns}}$. With these results, we find that $l(x) = C\left(-\frac{1}{2Ns}\right)e^{-2Nsx} + C^* = \left(-\frac{2Ns}{1-e^{-2Ns}}\right)\left(-\frac{1}{2Ns}\right)e^{-2Nsx} - \frac{e^{-2Ns}}{1-e^{-2Ns}} = \frac{e^{-2Nsx}-e^{-2Ns}}{1-e^{-2Ns}}$. It is easy to show that this result equals $1 - f(x)$.

Derivation 7

1. From the information given, which implies that the first subpopulation constitutes three-quarters of the population and the second one-quarter, $\bar{p}$ is immediately found to be $\frac{3 \cdot 0.7 + 1 \cdot 0.5}{4} = 0.65$, just as $V[p]$ is found to be $\frac{3 \cdot (0.7-0.65)^2 + 1 \cdot (0.5-0.65)^2}{4} = 0.0075$. In the first subpopulation, $H = 2 \cdot 0.7 \cdot (1 - 0.7) = 0.42$, and in the second subpopulation, $H = 2 \cdot 0.5 \cdot (1 - 0.5) = 0.50$, which implies that $H_S = \frac{3 \, 0.42 + 1 \cdot 0.50}{4} = 0.44$. From the value of $\bar{p}$ just calculated, we find that $H_T = 2 \cdot 0.65 \cdot (1 - 0.65) = 0.455$. Thus, $V[p]/\bar{p}(1 - \bar{p}) = 0.075/[0.65 \cdot (1 - 0.65)] \approx 0.03297$ and $(H_T - H_S)/H_T = (0.455 - 0.44)/0.455 \approx 0.03297$. Just as expected.

2. Since the frequency of the A_1B_1 haplotype is $0.7 \cdot 0.4 = 0.28$ in population I and $0.5 \cdot 0.1 = 0.05$ in population II, its frequency in the combined population is $(0.28 + 0.05)/2 = 0.165$. The frequency of allele A_1 in the combined population is equal to $(0.7 + 0.5)/2 = 0.6$, while the frequency of B_1 in this population is equal to $(0.4 + 0.1)/2 = 0.25$. Thus, according to (3.4a), we have that $D = 0.165 - 0.6 \cdot 0.25 = 0.015$.

3. The Wahlund effect is lost immediately with random mating in the combined population (disregarding the complexity with overlapping generations). Linkage disequilibria will, on the other hand, be lost more slowly, with a rate determined by the frequency of recombination between the associated loci. Even unlinked loci may show signs of the earlier population structuration for generations, taking into account that any linkage disequilibrium between such loci will decay at a rate of 50% – which is rapid but not instantaneous.

4. The frequencies of the six genotypes A_1A_1, A_2A_2, A_3A_3, A_1A_2, A_1A_3 and A_2A_3 will in the combined population be 0.085, 0.085, 0.34, 0.17, 0.16 and 0.16, while the same frequencies in a homogeneous population under Hardy–Weinberg assumptions would be 0.0625, 0.0625, 0.25, 0.125, 0.25 and 0.25. Comparing these values, one notes that all homozygotes in the combined population are more common than expected under Hardy–Weinberg and that (therefore) the sum of the heterozygotes is less common than expected. Perhaps surprising is, however, the frequency of the heterozygote genotype A_1A_2 in the combined population, 0.17, which is *in excess* of the expected Hardy–Weinberg value of 0.125. This result illustrates what is stated in the derivation, that even though the sum of heterozygotes will be smaller than expected under Hardy–Weinberg in a subdivided population, this does not necessarily hold for every possible heterozygous type.

5. The frequency of the allele in the large population is p. This will also be the mean of the allele frequencies in the subpopulations, but let us for clarity then denote it as $\bar{p}$. The variance in p among populations is given by the binomial variance $\bar{p}(1 - \bar{p})n/n^2$. Thus, among the created subpopulations, we find that $F_{ST} = V[p]/\bar{p}(1 - \bar{p}) = 1/n$. The chance-induced difference between these subpopulations becomes, thus, smaller with increasing sample size – just as expected. The result implies also that if we regard the original haploid population as consisting of individuals which each make up their own subpopulation, then F_{ST} becomes equal to 1. This is, undoubtedly, logical: A population cannot be thought of as being more subdivided than into the carriers of single gene copies.

Derivation 8

1. (a) In a sib-ship of mixed paternal origin, the offspring are normally less related than in the analysed situation of full sibs. Therefore, the donor of the warning must save a relatively higher number of sib-mates compared to the analysed case in order for this behaviour to spread when the mother has mated with many males. (b) Yes, it does matter – the evolution of danger-warning becomes easier. The

reason is most easily understood by considering the special case where *all* sib-mates are monozygotically related and therefore carry exactly the same genotype. Then, a behaviour where slightly more than one sib is saved per death due to warning will spread, compared to the more than two sibs required in the full-sib case. Dizygotic twins are genetically just like full sibs, so their frequency among the offspring does not affect the calculations in the derivation.

2. We start by considering what happens close to $p = 1$, which can be seen as the boundary equilibrium. If there are only a few A_2 gene copies in the population, then p is close to 1 and q is close to 0. Almost all of the A_2 alleles will be carried by heterozygotes (since random mating occurs in the population), and these will according to the assumptions have fitness $1 + a - bq$, which since q is close to 0 we can be approximate with $1 + a$. Since a is strictly positive, the rare heterozygotes A_1A_2 will have a higher fitness than the standard genotype, A_1A_1, with its fitness of 1. The A_2 allele will therefore always increase when rare, which is the same as saying that the $p = 1$ boundary equilibrium is unstable. But the A_1A_1 genotype does not always have the lowest fitness in the population. According to the assumptions, there is a state at $q = a/b$ where all genotypes have the same fitness. For all frequencies of A_2 above this value, the A_1A_1 homozygotes are most fit. The frequency of the A_2 allele, q, will therefore never increase above the value a/b. To sum up: The population will move to an internal equilibrium at a/b. This equilibrium is globally stable, even though all genotypes have exactly the same fitness at this point (there is, in other words, no direct heterozygote advantage here).

3. There are obviously $2313 \cdot 3011 \cdot 1985 \cdot 1335 \approx 1.846 \cdot 10^{13}$ possibilities of combining alleles at the different HLA loci described in the text, so this is the number of possible haplotypes for the chromosome stretch. In general, we know that a different haplotypes can be combined into $a \cdot (a + 1)/2$ diploid genotypes, which is close to $a^2/2$ when the number of haplotypes is high. Thus, there may be $(1.846^2/2) \cdot 10^{26} \approx 1.70 \cdot 10^{26}$ HLA genotypes in humans, all distinguishable by their different proteins. (Since there are additional protein-coding genes in the HLA region, many of which have a high number of variants, what we have calculated is, in fact, only a very

low estimate of the number of possible HLA genotypes. But with such a large number of possibilities, exactness hardly matters.)

4. The system described leads to a very simple logic of fertilization. To give but one example: S_1S_2 plants can only be fertilized by S_3 pollen, and the plants will produce equal frequencies of seeds with the S_1S_3 and S_2S_3 genotypes. Since we have assumed that there is a sufficient number of pollen of all kinds to fertilize all plants, the number of seeds produced by S_1S_2 plants will only depend on the number of such plants and not on the composition of the pollen cloud.

These considerations lead to the recurrence equations $a' = (b + c)/2$, $b' = (a + c)/2$ and $c' = (a + b)/2$. Starting with the initial genotype frequencies $a_0 = 0.9$, $b_0 = 0.1$ and $c_0 = 0$, we find that the genotypes in the next generation will have frequencies $a_1 = (0.1 + 0)/2 = 0.05$, $b' = (0.9 + 0)/2 = 0.45$ and $c_1 = (0.9 + 0.1)/2 = 0.5$. (Observe that the genotype lacking in generation 0 is now the most common one, while the genotype most common in generation 0 is now the least common one!) Continuing in the same way shows that $a_2 = (0.45 + 0.5)/2 = 0.475$, $b_2 = (0.05 + 0.5)/2 = 0.275$ and $c_2 = (0.05 + 0.45)/2 = 0.25$, and that $a_3 = (0.275 + 0.25)/2 = 0.2625$, $b_3 = (0.475 + 0.25)/2 = 0.3625$ and $c_3 = (0.475 + 0.45)/2 = 0.375$. An inspection of these frequencies, which in every generation become more equal to one another, plus the obvious symmetry in the recursion equations, make it reasonable to guess that the point $a = b = c = 1/3$, where the three alleles as well as the three genotypes are equally common, is an equilibrium to which the system will move. That this point is an equilibrium is easily shown by testing the recursion system with these values. We then find that $a' = (1/3 + 1/3)/2 = 1/3$, that $b' = (1/3 + 1/3)/2 = 1/3$ and that $c' = (1/3 + 1/3)/2 = 1/3$, showing that this point is indeed an equilibrium, since the recurrence equations transform the allele frequencies back to themselves.

Derivation 9

1. Irrespectively of the number of loci, the trait means in F_1 and F_2 are both 80 cm. This follows from the symmetry of the matings and the

additivity of the genetic effects involved. Now, for the variances:

There is in the F_1 population no genetic variance in height, since all plants have the same genotype. The phenotypic variance is then equal to the environmental variance alone: 8 cm^2.

To determine the variance in the F_2 population, we start by investigating what happens if only one locus is responsible for the genetic difference in height. Then, one-quarter of the population will be homozygous for the decreasing allele and have mean height 60 cm, one-half of the population will be heterozygous and have mean height 80 cm, and one-quarter of the population will be homozygous for the increasing allele and have mean height 100 cm. This makes the genetic variance in height (in cm^2) equal to:

$$\sigma_G^2 = E[X^2] - E[X]^2 = \tfrac{1}{4} \cdot 60^2 + \tfrac{1}{2} \cdot 80^2 + \tfrac{1}{4} \cdot 100^2 - 80^2 = 200.$$

Note that the variance only dependents on the *difference* in trait between the genotypes, and not on their specific numerical trait values. The subsequent calculations therefore become simpler if we let the value associated with the first line be 0 and the value associated with the second line be 40. We call this 'the transformed scale'.

We have, thus, for this special case, found that the total phenotypic variance (in cm^2) is $\sigma_P^2 = \sigma_G^2 + \sigma_E^2 = 200 + 8 = 208$. And, since all the genetic variance is additive ($\sigma_A^2 = \sigma_G^2$), the heritability is $h^2 = \sigma_A^2/\sigma_P^2 = 200/208 \approx 96\%$.

If not one but two loci are responsible for the difference in height between the two lines, then they must – according to the assumption of additivity over loci – cause half the difference each. The genetic variance generated by one of the loci is then (in cm^2) $\sigma_G^2 = \tfrac{1}{4} \cdot 0^2 + \tfrac{1}{2} \cdot 10^2 + \tfrac{1}{4} \cdot 20^2 - 10^2 = 50$. The total genetic variance (in cm^2) becomes, thus, twice this number (*i.e.* 100). This makes the phenotypic variance 108 cm^2 and the heritability $100/108 \approx 93\%$.

If we then go to k loci, the total genetic variance (still in cm^2) becomes $\sigma_G^2 = k \cdot \left[\tfrac{1}{4} \cdot 0^2 + \tfrac{1}{2} \cdot \left(\tfrac{20}{k}\right)^2 + \tfrac{1}{4} \cdot \left(\tfrac{40}{k}\right)^2 - \left(\tfrac{20}{k}\right)^2 \right] = k \cdot \left[\tfrac{1}{4} \cdot \left(\tfrac{40}{k}\right)^2 - \tfrac{1}{2} \cdot \left(\tfrac{20}{k}\right)^2 \right] = \tfrac{200}{k}.$

The genetic variance in F_2 depends, thus, directly on the inverse of the number of loci influencing the trait. The phenotypic variance in the trait is given by $(200/k) + 8$.

To summarize: The trait mean in the F_1 populations is 80 cm and the variance is 8 cm^2. In the F_2 population, the trait mean is 80 cm and the variance is $(200/k) + 8$ cm^2. In the F_2 population, we have that $\sigma_A^2 = \sigma_G^2 = 200/k$ cm^2, $\sigma_E^2 = 8$ cm^2, $\sigma_P^2 = (200/k) + 8$ cm^2 and $h^2 = \frac{200/k}{(200/k)+8} = 25/(25 + k)$.

2. The mean in the starting population is $k \cdot (0.6^2 \cdot 0/k + 2 \cdot 0.6 \cdot 0.4 \cdot 20/k + 0.4^2 \cdot 40/k) = k(16/k) = 16$ cm when the transformed scale introduced in Question 1 is used, or $60 + 16 = 76$ cm in the original scale.

After the described selective process, the new mean is $k \cdot (0.4^2 \cdot 0/k + 2 \cdot 0.6 \cdot 0.4 \cdot 20/k + 0.6^2 \cdot 40/k) = k(24/k) = 24$ cm in the transformed scale, or 84 cm in the original scale. Thus, the difference between the two means is 8 cm.

Since the genetic variance in the starting population is $\sigma_G^2 = k \cdot [0.6^2 \cdot (0/k)^2 + 2 \cdot 0.6 \cdot 0.4 \cdot (20/k)^2 + 0.4^2(40/k)^2] - (16/k)^2 = k(192/k^2) = 192/k$ cm^2, we have that $\sigma_G = \sqrt{192/k}$ cm. Note that it follows from symmetry considerations that the variance and standard deviation in the population take exactly the same values also *after* the assumed selection (you can easily check this yourself).

Thus, the result of the selection process, measured relative to the starting (and final) value for σ_G, equals $8/\sqrt{192/k} \approx 0.577 \cdot \sqrt{k}$. This implies that if the difference in height is genetically due to 4 segregating loci, then the result of selection becomes $1.15 \cdot \sigma_G$, while it becomes $5.77 \cdot \sigma_G$ if the difference is due to 100 segregating loci. In the first case, the overlap between the two distributions is high. In the second case, where the means differ by almost 6 standard variations, the overlap becomes negligible!

3. According to the results obtained in Question 1, the assumption that k equals 100 implies that $\sigma_A^2 = 2$ cm^2 and therefore that $\sigma_P^2 = \sigma_G^2 + \sigma_E^2 = \sigma_A^2 + \sigma_E^2 = 2 + 8 = 10$ cm^2. The strength of the described selection, S, is given as $0.1 \cdot \sigma_P$, which implies that $S = 0.1 \cdot \sqrt{10}$ cm. This makes the asked-for

response of selection, according to expression (9.2a), equal to
$R = \frac{\sigma_A^2}{\sigma_P^2} S = \frac{2}{10} \cdot 0.1 \cdot \sqrt{10} = \frac{0.2}{\sqrt{10}} \approx 0.063$ cm.

4. Following the notation used in the derivation, the mean difference between the two types of homozygotes at a segregating locus is $2a$. Thus, in this case (where $k = 100$), $a = 20/100 = 0.2$ cm. Since the selection coefficient at an individual locus according to expression (9.6a) is given by $s = \frac{2a}{\sigma_P^2} S$, we find that $s = \frac{2 \cdot 0.2}{10} \cdot 0.1 \cdot \sqrt{10} = \frac{0.04}{\sqrt{10}} \approx 0.013$ (where for S and σ_P^2 we utilize already obtained results).

When it comes to the effect of selection on the allele frequencies, we can use formula (1) for the change in allele frequency under selection from Background 2:3, where it is assumed that the fitness of the heterozygotes is intermediary to the fitness of the two types of homozygotes. According to this expression, and letting p be the frequency of the favoured allele, $p' = \frac{p(1 - sq/2)}{1 - sq}$. In the present situation, where selection acts on a previously unselected F_2 population, it is obvious that $p = q - 1/2$, which means that in the next generation the frequency of the increasing alleles will all be at $p' = \frac{1 - s/4}{2(1 - s/2)} = \frac{4 - s}{8 - 4s} \approx \frac{4 - 0.012649}{8 - 0.050596} \approx 0.5016$ (i.e. a frequency change of 0.0016).

Derivation 10

1. Let the proportion of $A_{sex}A_{apo}$ individuals be P and the proportion of $A_{sex}A_{sex}$ individuals be $1 - P$. Since only the $A_{sex}A_{apo}$ individuals produce pollen carrying the A_{apo} allele, the frequency of such pollen will be $P/2$.

(a) The following mating and offspring table is thereby obtained:

Mother plant	Fertilizing pollen	Frequency	Offspring seed	
			$A_{sex}A_{sex}$	$A_{apo}A_{sex}$
$A_{sex}A_{sex}$	A_{sex}	$(1 - P)(1 - P/2)$	1	0
$A_{sex}A_{sex}$	A_{apo}	$(1 - P)P/2$	0	1
$A_{sex}A_{apo}$	–	P	0	F

(b) The recursion equation for the frequency of apomictic plants becomes:

$$P' = [1 \cdot (1 - P)P/2 + F \cdot P]/[1 \cdot (1 - P) + F \cdot P] = P(1 - P + 2F)/2(1 - P + FP) = P(1 - P + 2F)/2[1 - P(1 - F)].$$

   To obtain this equation we have used the experience gained from deriving other such recursive systems in, for example, Derivations 8 and 10.

(c) When the apomictic plants have the same seed fertility as sexual plants (*i.e.* when $F = 1$), the recursion equation simplifies to $P' = P(3 - P)/2$. This implies that when P is small, $P' \approx 3P/2$, or in other words that the proportion of apomictic plants will increase by 50% every generation. This tremendously strong evolutionary effect derives from the fact that the apomictic plants – according to our exaggerated assumptions – can fertilize and produce off-spring via their male function as normal, at the same time as they directly copy the mother plant's genotype to the seed offspring and thereby avoid any genetic 'dilution' by unrelated pollen. Apomixis has then become like an infectious agent, transmitted both 'horizontally' via pollen and 'vertically' via the egg cell.

(d) Returning to the original recursion, it is seen that when the frequency of apomicts is small, the equation simplifies approximately to $P' \approx P(1 + 2F)/2$. Thus, the assumed mutation for apomixis will spread in the originally sexual population if $F > 1/2$, but will be lost from the population if $F < 1/2$. A more careful study of the recursion equation shows that no internal equilibrium is possible, which implies that if the mutation spreads when rare, then it will continue to do so until apomixis has become the fixed breeding system in the population.

2. In the adult breeding population, there will only be individuals with no or one B chromosome. Let the frequency of those with one B chromosome be P; this frequency will be the same in males and females.

(a) The mating/offspring table is as follows:

Female	Male	Frequency	Offspring		
			Without B	With 1 B	With 2 B
Without B	Without B	$(1 - P)(1 - P)$	1	0	0
Without B	With 1 B	$(1 - P)P$	$1/2$	$1/2$	0
With 1 B	Without B	$P(1 - P)$	$1 - d$	d	0
With 1 B	With 1 B	P^2	$1/2(1 - d)$	$1/2$	$1/2(d)$

(b) The recursion for the frequency of animals with one B chromosome will be $P' = [(1 - P) \cdot P/2 + P \cdot (1 - P) \cdot d + P^2/2]/[1 - P^2 \cdot d/2]$, when the lethal effect of inheriting two B chromosomes is accounted for.

(c) For small values on P, this equation simplifies to $P' \approx P/2 + Pd = P(d + 1/2)$, which shows that the described B chromosome will spread when rare if $d > 1/2$ (*i.e.* as soon as it is favoured by any segregation distortion). And this despite the fact that its only phenotypic effect, besides pushing for itself during female meiosis, is deleterious.

(d) At a valid internal equilibrium (*i.e.* a state where $0 < P < 1$), the frequency P must be such that $P = [(1 - P) \cdot P/2 + P \cdot (1 - P) \cdot d + P^2/2]/[1 - P^2 \cdot d/2]$, which can be easily simplified to $P = 1 - \sqrt{(1 - d)/d}$. For $d > 1/2$, there is, thus, always an internal equilibrium to the process. Without further analysis, we accept that it is to this value for P that the population will move when the B chromosome is favoured by segregation distortion.

3. (a) Let the proportion of plants with the CMS cytoplasm be P. According to the assumptions in the question, such plants will produce F seeds relative to other plants. Thus, we get that

$P' = P \cdot F/[P \cdot F + (1 - P) \cdot 1] = P\frac{F}{1+P(F-1)}$. For small values on P, the recursion approximates to $P' \approx FP$, which implies that the CMS cytoplasm will spread if $F > 1$. To investigate the equilibria to this process, we set $P = P\frac{F}{1+P(F-1)}$, which (for $F \neq 1$) is solved by either $P = 0$ or $P = \frac{F-1}{F-1} = 1$. The only equilibria to the process are therefore the two boundary equilibria, $P = 0$ and $P = 1$. The result also implies that if the first equilibrium ($P = 0$) is unstable, then the second ($P = 1$) must be stable.

(b) According to the results just obtained, a CMS mutation which spreads in a population due to $F > 1$ will ultimately go to fixation. This is, however, absurd, since all plants in the population will then be male sterile, implying that no functional pollen is produced – and how then are the plants fertilized and the seeds produced? The only possibility is therefore that the population goes extinct at the same time as the CMS cytoplasm becomes fixed.

(c) Yes, one must realise that it is perfectly possible that a population goes extinct due to the spread and fixation of some unfortunate mutation – this follows from the blind functioning of evolution. But in real life, other evolutionary scenarios are also worth considering.

It is, for example, possible that the assumed advantage in seed production of the CMS plants (the assumed fixed fitness advantage, $F - 1$) in fact disappears when the amount of pollen in the population decreases. The question states that the plant is outbreeding, but if self-fertilizations are not forbidden then the seed production of normal plants may well be smaller than the seed production of CMS plants when these are rare. However, this effect will disappear and change into an advantage for normal plants when pollen starts to become scarce in the population. In a more detailed model, F would then be treated as a function that decreases with the frequency, P. In such a situation, one would expect that the population would move to a polymorphic state with both male sterile and male fertile plants present.

A third possibility is that a nuclear mutation that restores at least some pollen production to otherwise male sterile plants appears. Maybe its other fitness effects are such that it would not normally be favoured in evolution, but in a situation with many male sterile plants and almost no pollen produced, such a mutation will be strongly favoured. The outcome may again be a population polymorphic for male sterile and male fertile plants. Nuclear 'restorers' of the assumed kind are well known from many plant species in which CMS occurs.

We have here encountered a situation where the first, simple, population genetic analysis leads to a result which prompts deeper studies and further modelling of the evolutionary possibilities available. We are thereby reminded that no analysis gives results that can be taken as definitive and fixed. Life is always good at inventing new and sometimes highly unexpected alternatives, and there are always new models to be proposed and studied, to see what insights they lead to.

REFERENCES

For the names of all authors of articles with five or more contributors, please look at the originals.

Balding, J. D., Bishop, M. and Cannings, C. 2007. *Handbook of Statistical Genetics*, 3rd edn. John Wiley & Sons.

Barton, N. and Bengtsson, B. O. 1986. The barrier to genetic exchange between hybridising populations. *Heredity* **56**: 357–376.

Bengtsson, B. O. 1985. The flow of genes through a genetic barrier. In *Evolution – Essays in Honour of John Maynard Smith* (Greenwood, P. J., Harvey, P. H. and Slatkin, M., eds), pp. 31–42. Cambridge University Press.

Bersaglieri, T., Sabeti, P. C., Patterson, N., *et al.* 2004. Genetic signatures of strong recent positive selection at the lactase gene. *Am. J. Hum. Genet.* **74**: 1111–1120.

Bodmer, W. F. 1965. Differential fertility in population genetics models. *Genetics* **51**: 411–424.

Bodmer, W. F. and Cavalli-Sforza, L. L. 1968. A migration matrix model for the study of random genetic drift. *Genetics* **59**: 565–592.

Burt, A. and Trivers, R. 2008. *Genes in Conflict. The Biology of Selfish Genetic Elements*. Belknap Press.

Cavalli-Sforza, L. L. and Bodmer, W. F. 1971. *The Genetics of Human Populations*. Freeman.

Charlesworth, B. 1994. *Evolution in Age-Structured Populations*, 2nd edn. Cambridge University Press.

Charlesworth, B., Morgan, M. T. and Charlesworth, D. 1993. The effect of deleterious mutations on neutral molecular variation. *Genetics* **134**: 1289–1303.

Christiansen, F. B. 1991. On conditions for evolutionary stability for a continuously varying character. *Amer. Naturalist* **138**: 37–50.

Christiansen, F. B. 2008. *Theories of Population Variation in Genes and Genomes*. Princeton University Press.

Understanding Population Genetics, First Edition. Torbjörn Säll and Bengt O. Bengtsson.
© 2017 John Wiley & Sons Ltd. Published 2017 by John Wiley & Sons Ltd.

References

Christiansen, F. B. and Fenchel, T. M. 1977. *Theories of Populations in Biological Communities*. Springer.

Christiansen, F. B. and Loeschcke, V. 1980. Evolution and intraspecific exploitative competition. I. One-locus theory for small additive gene effects. *Theor. Popul. Biol.* **18**: 297–313.

Corbett-Detig, R. B., Hartl, D. L. and Sackton, T. B. 2015. Natural selection constrains neutral diversity across a wide range of species. *PLOS Biology* DOI:10.1371/journal.pbiol.1002112.

Crow, J. F. and Kimura, M. 1970. *An Introduction to Population Genetics Theory*. Harper & Row.

Darwin, C. 1859. *On the Origin of Species by Means of Natural Selection, or the Preservation of Favoured Races in the Struggle for Life*. Murray. [Many later editions and reprints.]

Darwin, C. 1871. *The Descent of Man and Selection in Relation to Sex*. Murray. [Many later editions and reprints.]

Dedekind, R. 1888. *Was sind und was sollen die Zahlen?* Vieweg, Braunschweig. [Translation: Dedekind, R. 1963. *Essays on the Theory of Numbers*. Dover.]

Dunn, L. C. 1953. Variations in the segregation ratio as causes of variations in gene frequencies. *Acta Genet.* **4**: 139–147.

Edwards, A. W. F. 2000. *Foundations of Mathematical Genetics*, 2nd edn. Cambridge University Press.

Ellegren, H., Smeds, L., Burri, R., *et al.* 2012. The genomic landscape of species divergence in *Ficedula* flycatchers. *Nature* **491**: 756–760.

Ewens, W. J. 1979. *Mathematical Population Genetics*. Springer.

Falconer, D. S. 1965. The inheritance of liability to certain diseases, estimated from the incidence among relatives. *Ann. Hum. Genet.* **29**: 51–76.

Falconer, D. S. and Mackay, T. F. C. 1996. *Introduction to Quantitative Genetics*, 4th edn. Longman.

Feldman, M. W. 1972. Selection for linkage modification: I. Random mating populations. *Theor. Popul. Biol.* **3**: 324–346.

Feldman, M. W., Otto, S. P. and Christiansen, F. B. 1997. Population genetic perspectives on the evolution of recombination. *Annu. Rev. Genet.* **30**: 261–295.

Felsenstein, J. 2004. *Inferring Phylogenies*. Sinauer.

Fisher, R. A. 1918. The correlation between relatives on the supposition of Mendelian inheritance. *Trans. Roy. Soc. Edinb.* **52**: 399–433.

Fisher, R. A. 1922. On the dominance ratio. *Proc. Roy. Soc. Edinb.* **42**: 321–341.

Fisher, R. A. 1930. *The Genetical Theory of Natural Selection*. Clarendon Press.

Gershenson, S. 1928. A new sex-ratio abnormality in *Drosophila obscura*. *Genetics* **13**: 488–507.

Gillespie, J. H. 2004. *Population Genetics. A Concise Guide*, 2nd edn. Johns Hopkins University Press.

Haldane, J. B. S. 1924. A mathematical theory of natural and artificial selection. Part I. *Trans. Camb. Phil. Soc.* **23**: 19–41.

Haldane, J. B. S. 1927. A mathematical theory of natural and artificial selection. Part V. Selection and mutation. *Proc. Camb. Phil. Soc.* **23**: 838–844.

Haldane, J. B. S. 1955. Population genetics. *New Biol.* **18**: 34–51.

Hamilton, W. D. 1963. The evolution of altruistic behaviour. *Amer. Naturalist* **97**: 354–356.

Hamilton, W. D. 1967. Extraordinary sex ratios. *Science* **156**: 477–488.

Hardy, G. H. 1908. Mendelian proportions in a mixed population. *Science* **28**: 49–50.

Hedrick, P. W. 2011a. *Genetics of Populations*, 4th edn. Jones & Bartlett.

Hedrick, P. W. 2011b. Population genetics of malaria resistance in humans. *Heredity* **107**: 283–304.

Hein, J., Schierup, M. H. and Wiuf, C. 2005. *Gene Genealogies, Variation and Evolution*. Oxford University Press.

Hill, W. G., Goddard, M. E. and Visscher, P. M. 2008. Data and theory point to mainly additive genetic variance for complex traits. *PLOS Genetics* **4**: e1000008.

Hurst, L. D., Atlan, A. and Bengtsson, B. O. 1996. Genetic conflicts. *Quart. Rev. Biol.* **71**: 317–364.

Keightley, P. D. and Otto, S. P. 2006. Interference among deleterious mutations favours sex and recombination in finite populations. *Nature* **443**: 89–92.

Kimura, M. 1962. On the probability of fixation of mutant genes in a population. *Genetics* **47**: 713–719.

Kimura, M. 1968. Evolutionary rate at the molecular level. *Nature* **217**: 624–626.

Kimura, M. 1969. The number of heterozygous nucleotide sites maintained in a finite population due to steady flux of mutations. *Genetics* **61**: 893–903.

Kimura, M. 1980. A simple method for estimating evolutionary rates of base substitutions through comparative studies of nucleotide sequences. *J. Mol. Evol.* **16**: 111–120.

Kimura, M. 1983. *The Neutral Theory of Molecular Evolution*. Cambridge University Press.

Kimura, M. and Crow, J. F. 1964. The number of alleles that can be maintained in a finite population. *Genetics* **49**: 725–738.

Kimura, M. and Crow, J. F. 1978. Effect of overall phenotypic selection on genetic change at individual loci. *Proc. Natl. Acad. Sci. USA* **75**: 6168–6171.

Kimura, M. and Weiss, G. H. 1964. The stepping stone model of population structure and the decrease of genetic correlation by distance. *Genetics* **49**: 561–576.

Kingman, J. F. C. 1961. On an inequality in partial averages. *Quart. J. Math.* **12**: 78–80.

Kingman, J. F. C. 1982. On the genealogy of large populations. *J. Appl. Prob.* **19A**: 27–43.

Kuhn, T. S. 1977. A function for thought experiments. In *The Essential Tension: Selected Studies in Scientific Tradition and Change*, pp. 240–265. University of Chicago Press.

Lakatos, I. 1976. *Proofs and Refutations. The Logic of Mathematical Discovery.* Cambridge University Press.

Lewis, D. 1941. Male sterility in natural populations of hermaphroditic plants. *New Phytologist* **40**: 56–63.

Lewontin, R. C. 1974a. *The Genetic Basis of Evolutionary Change.* Columbia University Press.

Lewontin, R. C. 1974b. The analysis of variance and the analysis of causes. *Am. J. Hum. Genet.* **26**: 400–411.

Lewontin, R. C. and Hubby, J. L. 1966. A molecular approach to the study of genetic heterozygosity in natural populations. II. Amount of variation and degree of heterozygosity in natural populations of *Drosophila pseudoobscura*. *Genetics* **54**: 595–609.

Malécot, G. 1948. *Les mathématiques de l'hérédité*. Masson. [Reprinted in Malécot (1966). Translated as: Malécot, G. 1969. *The Mathematics of Heredity*. Freeman.]

Malécot, G. 1966. *Probabilité et hérédité*. Presses Universitaires de France.

Mandel, S. P. H. and Hughes, I. M. 1958. Change in mean viability at a multi-allelic locus in a population under random mating. *Nature* **182**: 63–64.

Manolio, T. A., Collins, F. S., Cox, N. J., *et al.* 2009. Finding the missing heritability of complex diseases. *Nature* **461**: 747–753.

Marais, G. 2003. Biased gene conversion: implications for genome and sex evolution. *Trends Genet.* **19**: 330–338.

Maynard Smith, J. 1964. Group selection and kin selection. *Nature* **201**: 1145–1147.

Maynard Smith, J. and Haigh, J. 1974. The hitch-hiking effect of a favourable gene. *Genet. Res.* **23**: 23–35.

Maynard Smith, J. and Price, G. R. 1973. The logic of animal conflict. *Nature* **246**: 15–18.

McQuillan, R., Leutenegger, A. L., Abdel-Rahman, R., *et al.* 2008. Runs of homozygosity in European populations. *Am. J. Hum. Genet.* **83**: 359–372.

Mead, S., Whitfield, J., Poulter, M., *et al.* 2008. Genetic susceptibility, evolution and the kuru epidemic. *Phil. Trans. Roy. Soc. B* **363**: 3741–3746.

Nei, M. 1967. Modification of linkage intensity by natural selection. *Genetics* **57**: 625–641.

Nei, M., Maruyama, T. and Chakraborty, R. 1975. The bottleneck effect and genetic variability in populations. *Evolution* **29**: 1–10.

Nilsson-Ehle, H. 1909. Kreutzunguntersuchungen an Hafer und Weizen. *Lunds Universitets Årsskrift*, N. F. Afd. 2, Bd. 5, No. 2.

Ohta, T. 1973. Slightly deleterious mutant substitutions in evolution. *Nature* **246**: 96–98.

Östergren, G. 1945. Parasitic nature of extra fragment chromosomes. *Botaniska Notiser*, Häfte 2, 157–163.

Poelstra, J. W., Vijay, N., Cossu, C. M., *et al.* 2014. The genomic landscape underlying phenotypic integrity in the face of gene flow in crows. *Science* **344**: 1410–1414.

Prado-Martinez, J., Sudmant, P. H., Kidd, J. M., *et al.* 2013. Great ape genetic diversity and population history. *Nature* **499**: 471–475.

Price, G. R. 1970. Selection and covariance. *Nature* **227**: 520–521.

Provine, W. B. 1971. *The Origins of Theoretical Population Genetics*. University of Chicago Press.

Robbins, R. B. 1918. Some applications of mathematics to breeding problems. III. *Genetics* **3**: 375–389.

Robertson, A. 1960. A theory of limits in artificial selection. *Proc. Roy. Soc. B* **153**: 234–249.

Robinson, J., Halliwell, J. A., Hayhurst, J. D., *et al.* 2015. The IPD and IMGT/HLA database: allele variant databases. *Nucl. Acids Res.* **43**: D423–D431.

Säll, T., Jakobsson, M., Lind-Halldén, C. and Halldén, C. 2003. Chloroplast DNA indicates a single origin of the allotetraploid *Arabidopsis suecica. J. Evol. Biol.* **16**: 1019–1029.

Scally, A., Dutheil, J. Y., Hillier, L. W., *et al.* 2012. Insights into hominid evolution from the gorilla genome sequence. *Nature* **483**: 169–175.

Schlebusch, C. M., Sjödin, P., Skoglund, P. and Jakobsson, M. 2013. Stronger signal of recent selection for lactase persistence in Maasai than in Europeans. *Eur. J. Hum. Genet.* **21**: 550–553.

Solomon, E. and Bodmer, W. F. 1979. Evolution of sickle variant gene. *Lancet* **313**: 923.

References

Tajima, F. 1989. Statistical method for testing the neutral mutation hypothesis by DNA polymorphism. *Genetics* **123**: 585–595.

Uyenoyama, M. K. and Bengtsson, B. O. 1979. Towards a genetic theory for the evolution of the sex ratio. *Genetics* **93**: 721–736.

Uyenoyama, M. and Feldman, M. W. 1980. Theories of kin and group selection: a population genetics perspective. *Theor. Popul. Genet.* **17**: 380–414.

Wahlund, S. 1928. Zusammensetzung von Populationen und Korrelations- erscheinungen vom Standpunkt der Vererbungslehre aus betrachtet. *Hereditas* **11**: 65–106. [Translated as: Wahlund, S. 1975. Composition of populations and of genotypic correlations from the viewpoint of population genetics. In *Demographic Genetics* (Weiss, K. M. and Ballonoff, P. A., eds), pp. 224–263. Dowden, Hutchinson & Ross.]

Weinberg, W. 1908. Über den Nachweis der Vererbung beim Menschen. *Jahresh. Verein. f. vaterl. Naturk. Württem.* **64**: 369–382.

Weir, B. S. 1996. *Genetic Data Analysis II: Methods for Discrete Population Genetic Data*. Sinauer.

Williams, G. C. and Williams, D. C. 1957. Natural selection of individually harmful social adaptations among sibs with special reference to social insects. *Evolution* **11**: 32–39.

Wright, S. 1931. Evolution in Mendelian populations. *Genetics* **16**: 97–159.

Wright, S. 1937. The distribution of gene frequencies in populations. *Proc. Natl. Acad. Sci. USA* **23**: 307–320.

Wright, S. 1943. Isolation by distance. *Genetics* **28**: 114–138.

Wright, S. 1951. The genetical structure of populations. *Ann. Eugen.* **15**: 323–354.

INDEX

Understanding Population Genetics, First Edition. Torbjörn Säll and Bengt O. Bengtsson.
© 2017 John Wiley & Sons Ltd. Published 2017 by John Wiley & Sons Ltd.